T0214966

Embracing Indigenous Knowledge in Science and Medical Teaching

Cultural Studies of Science Education
Volume 10

Series Editors
KENNETH TOBIN, *City University of New York, USA*
CATHERINE MILNE, *New York University, USA*
CHRISTINA SIRY, *University of Luxembourg, Walferdange, Luxembourg*
MICHAEL P. MUELLER, *Anchorage, Alaska, USA*

The series is unique in focusing on the publication of scholarly works that employ social and cultural perspectives as foundations for research and other scholarly activities in the three fields implied in its title: science education, education, and social studies of science.

The aim of the series is to establish bridges to related fields, such as those concerned with the social studies of science, public understanding of science, science/technology and human values, or science and literacy. *Cultural Studies of Science Education*, the book series explicitly aims at establishing such bridges and at building new communities at the interface of currently distinct discourses. In this way, the current almost exclusive focus on science education on school learning would be expanded becoming instead a focus on science education as a cultural, cross-age, cross-class, and cross-disciplinary phenomenon.

The book series is conceived as a parallel to the journal *Cultural Studies of Science Education*, opening up avenues for publishing works that do not fit into the limited amount of space and topics that can be covered within the same text.

More information about this series at http://www.springer.com/series/8286

Mariana G. Hewson

Embracing Indigenous Knowledge in Science and Medical Teaching

 Springer

Mariana G. Hewson
Center for Medical Education Research and Development
The Cleveland Clinic Foundation (Retired)
Madison, WI, USA

Additional material to this book can be downloaded from http://extras.springer.com

ISSN 1879-7229   ISSN 1879-7237 (electronic)
ISBN 978-94-024-0524-8   ISBN 978-94-017-9300-1 (eBook)
DOI 10.1007/978-94-017-9300-1
Springer Dordrecht Heidelberg New York London

Printed on acid-free paper

Springer is part of Springer Science+Business Media (www.springer.com)

For Jim Posiwa Mudau

For three decades, you were part of our family in Johannesburg. All four of us children looked to you as our second father. While you worked, you also influenced our upbringing by telling us African stories, riddles, and folklore. You insisted on appropriate politeness at all times. You cooked for all our celebrations, baked our birthday cakes, came with us on our family seaside holidays, and you were present at our weddings.

I remember you once told me, "Go and do your homework so you can be a doctor." I wonder what kind of doctor you had in mind. Did you ever know how much that meant to me?

Through knowing you I developed a great love for, and appreciation of, African people and their ways. This book is the result of your profound influence on my life.

Acknowledgments

I wish to express grateful thanks to all the people involved in this project. I am in awe of the amount of support, love, and cherishing that it has taken to write a book.

In the Western Cape studies, I am deeply grateful to Mirranda T. Javu, traditional healer, who provided me invaluable connections with traditional healers in the Western Cape. With charm and vitality, she facilitated all my meetings with traditional healers in Cape Town and the surrounding areas. She has become a colleague and a dear friend. I also thank Fadli Wagiet, Curriculum Advisor, Western Cape Education Department, for helpful suggestions and for recruiting the group of expert science teachers to respond to the ideas of the traditional healers.

In the Lesotho study, I thank Dr. Maseqobela Williams, Director of the Department of Science and Technology, Maseru, Lesotho, who agreed to support my project. She appointed T'sepang Ledia in her department to help me throughout my visit. He came with me to all the sessions with the traditional healers and served as my interpreter whenever needed. His lively interest in the study and his warmth and humor helped immensely. Dr. Lira Molapo, Science Advisor from the Department of Education, Lesotho, generously shared his viewpoint on the state of science education in Lesotho. I sincerely thank Linus Liau, head of the Traditional Healers Association, Lesotho, who made it possible for me to meet the senior leaders of traditional healers as well as the two large groups of healers in Lesotho. My special thanks go to these traditional healers who were willing to attend the sessions (sometimes walking vast distances across mountains) to share their extensive knowledge about traditional African healing.

Christopher Lang, videographer from Madison, WI, volunteered to accompany me to Lesotho and record everything. I greatly appreciate his talent in producing the videotape.

I am deeply grateful to several traditional healers in southern Africa. In Maputo, Mozambique, I met Juliette and Celeste. We talked about traditional healing, and they showed me how to help initiates in their training. They were happy to have the conversations with me and honored me with the offer of a live chicken to take home to the United States.

Elizabeth Mogari, who lives in Rustenburg, South Africa, welcomed me into her home and her confidence and taught me a great deal about healing. She allowed me to be a participant observer in several healing sessions—a priceless gift. Evelyn Maliti, a traditional healer in Zwelihle in the Western Cape, showed me so much about healing practices as well as her way of training apprentices. In Bulawayo, Zimbabwe, Mandaza Augustine Kandemwa accepted me as an initiate in his training program. I have greatly appreciated his open-hearted willingness to initiate me into the healing arts of Africa. Mandaza is a born teacher and a wise man.

My student and now my colleague, Kamonwan Kanyaprasith, Ph.D., in Bangkok, Thailand (science educator), provided me with details about Nayao School in Thailand. Janine C. Edwards, Ph.D. (medical educator), talked with me about her views on how westerners enter medical training. She also critically read the first draft of the manuscript. Two physician colleagues, Rebecca Byers (general internist from the University of Wisconsin, Madison) and Jeffery Hutzler (psychiatrist from the Cleveland Clinic Foundation, Cleveland, Ohio), accompanied me on separate visits to South Africa and helped me interpret the interactions from a medical point of view. Jean Hutzler (Director of Public Policy, Catholic Charities Health and Human Services, Cleveland, Ohio) helped with recording some of the interviews.

I sincerely thank several people for giving me interviews that I have recounted (in part) in this book. Meshach Ogunniyi, Ph.D., Professor of Science Education at the University of the Western Cape, provided me with funds to help in making the videotape that accompanies this book and offered me collegiality and intellectual support for my independent research projects in recent years.

As I think about the people who have affected me during my life, many come to mind. Among these are: Ella Le Maitre (Headmistress, Roedean School, Johannesburg), John Turner (Professor of Education, University of Lesotho, Botswana and Swaziland), Joe Novak (Professor of Science Education, Cornell University), Alan Knox (Professor of Education, University of Wisconsin), Thomas C. Meyer (cardiologist and medical educator, University of Wisconsin), and many colleagues in the American Academy on Communication in Healthcare. You have mentored me in my career and given me your trust as well as confidence in myself. You have shared your insights, passions, beliefs, knowledge, and skills. You continue to inspire me always—that's what spirits do.

My support team at home has been enormously encouraging. Janine Edwards, Ph.D. (medical educator), Adele Gordon, Ph.D. (mathematics educator), Mike Goldblatt (photographer), Karen Kendrick-Hands, J.D., and so many others have critically read drafts and/or watched the making of the video and generally cheered me on. Peter Hewson, D.Phil., and Katharine Hewson, M.A., took the beautiful photographs.

I also thank Ken Tobin, Springer Series Editor, for courageously inviting me to write this book. Bernadette Ohmer of Springer Science+Business Media in Holland frequently gave me calm assurance that despite all the vicissitudes in life, all would be well. And it was my husband Peter Hewson supported me in this project from the start. He sanguinely waved me off on my trips to visit the traditional healers in Zimbabwe, Mozambique, Lesotho, and various parts of South Africa without knowing exactly where I was going or what I was planning to do when I got there. Without his complete confidence, enduring love and steady presence I could not have accomplished this project.

Contents

Biography

In between climbing mountains, traveling the world, getting married, and raising a family, I studied. My Bachelor of Science degree came from the University of the Witwatersrand, Johannesburg. I studied at Oxford University, UK, for my Diploma in Education. I earned my Master in Science Education from the University of British Columbia, Canada, and my Doctorate in Education from the University of the Witwatersrand, Johannesburg—back where I started. For the next 10 years I became a science educator, pursuing the same line of research that I had worked on in my doctoral study. The highlight of this period was the opportunity to teach science education at the University of Botswana, Lesotho, and Swaziland. During that time, I had my first meeting with a Basotho traditional healer and have had a sustained academic interest in their roles, training, and views ever since.

My involvement with medical education began in the late 1980s after we emigrated from South Africa to the United States. Medical education became the focus of my career for the next 20 years. Of great significance in my career, I was awarded the title of Fellow of the American Academy on Communication in Healthcare. For many years, I participated in the work of improving the quality and effectiveness of healing in western healthcare through improved doctor-patient communication. I culminated my career as Director of Medical Education Research and Development at the Cleveland Clinic, USA—a precious time of fulfilling work. During that period, I also became interested in Complementary and Alternative Medicine (CAM). With my colleague Joan Fox, we organized a faculty development program for introducing CAM to physicians. This involved embracing the diverse views of western physicians and non-western medical practitioners.

After retiring, I pursued my interests in indigenous African medical practices by interviewing traditional healers during my visits to South Africa. Several times I was invited by these healers to become an initiate in the African practice. They told me that if I was so interested in their way of practicing medicine and their indigenous knowledge, it must mean that my own spirits were calling me to the practice.

This put me in an awkward situation—how could I, as a scientist and science educator, get involved with an indigenous training process? As a westerner, I felt that it was impossible to operate both in the western way of making sense of the world and also the African way. I felt that if I did go through the initiation process, I would be betraying both my Christian and scientific commitments.

My narrative on my initiation as a healer (see Chap. 10) shows the difficulties I faced, both emotionally and intellectually. My experiences as a scientifically trained westerner can be no more challenging than of traditional Africans trying to learn the western view of science. I was disconcerted by the centrality of spirituality in the African understanding of the world. For Africans, I imagine that learning western science feels disconcerting because of the central stance of objectivity and the lack of spirituality.

Part I
Introduction

Chapter 1
Prologue

Abstract I discuss different ways of knowing as conceived by western and indigenous people. I set out to find ways to build bridges between these different ways in order to improve teaching and learning in Africa. I outline the structure of the book and explain my use of different writing approaches.

Mountain Meeting, South Africa, 2006

I was seeking a traditional healer in Kwa-Zulu Natal who would be willing to talk with me. So I asked a man in Underberg where I could find a traditional healer, and he pointed the way to Daisy's home.

Daisy is a traditional healer who lives in the countryside in a metal rail container. Her tiny home is located in the foothills of the Drakensberg, the high mountain ridge that runs north to south in South Africa. The peaks and ridges were topped with snow, causing a chilly, dry wind down below. The grass was long and brown, contrasting with the blue winter sky.

As luck would have it, Daisy was at home. I explained my mission, to learn about traditional healers and their work, for the benefit of all healers. We had a good talk about her life as a healer, her beliefs, and her healing practices. She mostly confirmed things I already knew from other traditional healers in different parts of the country. Then I wondered aloud why she was so willing to talk to me. Daisy told me that if a person knows about the spirits and healing powers and lives her life closely connected with the spirits, and if a stranger genuinely wants to learn about this, then that healer MUST share her knowledge. If a traditional healer withholds this information, that is so central to the physical and spiritual well-being of humans, the spirits will punish him or her. They will send misfortunes such as the loss of a child, a house burned down, a terrible illness, and possible death. I got it! Daisy was instructing me, enlightening me, and showing me that the price of special knowledge is to share it with others.

In this book I share my knowledge about traditional healers in southern Africa and their healing approaches that I have accumulated over the last 35 years. I outline my own journey of coming to understand and embrace the knowledge and skills of

© Springer Science+Business Media Dordrecht 2015

M.G. Hewson, *Embracing Indigenous Knowledge in Science and Medical Teaching*, Cultural Studies of Science Education 10, DOI 10.1007/978-94-017-9300-1_1 traditional healers. It has not been a straightforward process. As a westerner, I trained in science, the teaching of science, as well as the teaching of medicine. I describe my own doubts and confusions about bridging the interface between western science and the African indigenous knowledge, particularly that of traditional healers in southern Africa.

My intention is to bring the indigenous knowledge of Africa out of the murky shadows into the light for the benefit of science and medical education. The problems that face our world today are large and troubling. Will there be a world for our grandchildren and beyond, or will we have destroyed it beyond recognition, beyond habitation? What can the leaders of the world do about problems such as global warming, deforestation, soil erosion, loss of diversity of plant and animal species, unsustainable macro farming techniques, harmful chemicals that pollute our environments—land, sea and water, raging diseases (some of which are caused by ourselves), and a descent into shooting innocent, vulnerable people (such as small children at schools and universities, people at religious services, athletes running a marathon, people enjoying a film) for no particular reasons.

I have three beliefs about solutions to this dysfunction. One of these is that the knowledge, attitudes and skills of indigenous people might hold the answers, at least in part, to the world's current problems. I also believe that until indigenous knowledge becomes a part of the curriculums in educational institutions we will not be able to remediate the learning difficulties of non-westerners. And third, I take the position that multiple ways of knowing (epistemological pluralism) is a valid approach to solving complex problems.

This book is intended to introduce readers to ways of knowing that differ from the western way. My basic idea is for westerners to step out of their western shoes and go barefoot for a while in order to experience a different way of knowing about, understanding, and approaching the complexities of the world that we all share.

The focus of this book is on teaching science and medicine in two worlds: western and non-western, namely African. I distinguish between the indigenous knowledge of non-westerners, and the scientific and medical knowledge of westerners. By highlighting the differences between these two ways of knowing I seek to elucidate a basic reason for the dramatic lack of success of African children in science topics, and confusions in medical care giving and teaching. My intention is to find ways to bridge the cultural divide between western and non-western ways of knowing, especially concerning the teaching of science and the practice and teaching of medicine.

Writing Styles

In this book I use four different writing styles. One is regular academic writing that offers text with references. Then I also include narratives, interviews and autoethnography.

Narrative. I use narratives to present ideas, conflicts between ideas, and occasionally, resolution of conflicts. Stories provide depth, drama, and emotion, and promote reflection. They are useful for exploring complexity, uncovering assumptions and biases, and eliciting different perspectives. Stories make the content of a text more memorable than the hard factual knowledge, and language of academia (Mullan et al. 2006; Hatem 2014).

Polyphony. My approach also involves multiple voices. I provide several interviews with people who contribute their views on the discontinuities between western and indigenous thought patterns, and also the differences in thinking skills in both science and medical teaching. I allow multiple voices to illuminate the differences between thinking and knowing in different cultural settings. This allows me to weave a complex view of the problem.

Autoethnography. This approach is close to the arts and humanities. It allows for subjectivity and is committed to cultural questioning and analysis and is therefore different from biography or memoir. Autoethnography reveals multi-layered levels of reflection about complex phenomena involved in teaching in two epistemologically different worlds, e.g., teachers in multi-cultural science classrooms, western teachers who teach homogenous students coming from a different knowledge system, and *vice versa.*

> (A)utoethnography is a contemporary qualitative research methodology, demanding unusually rigorous, multi-layered levels of researcher reflexivity, given that the researcher/s and the researched are normally the same people. (This involves) research, writing, and methods that connect the autobiographical and personal to the cultural and social. The form usually features concrete action, emotion, embodiment, self-consciousness, and introspection … (and) claims the conventions of literary writing. (Ellis et al. 2008; Short et al. 2013)

To reflect on my personal journey (autoethnography) I recount stories from my own experiences. I also recount the stories of others. Both the stories and interviews are presented in text-boxes to distinguish the more literary voices from the academic text.

Scope of the Book

This book has five parts. In Part I, I introduce the main concerns and ideas that have occupied me during my careers in both science and medical education. Part II focuses in more depth on different ways of knowing, especially in the west and southern Africa. In Part III, I look at the challenges of different ways of knowing in medicine, especially concerning medical practitioners teaching their patients and their medical students. Part IV is concerned with implications of these ideas for teaching both science and medicine. Finally, Part V describes my personal journey in understanding the African way of knowing about medicine through my initiation in Zimbabwe. I conclude with a video made during my research in Lesotho in which traditional healers told me about their ideas concerning the educational needs of Basotho children.

There are four conceptual components and two main strands to this teaching. The four components are western knowledge, indigenous knowledge, science teaching, and medical teaching. By dealing with both school science and medical education, I gain a wide and varied palette to make my points—after all, medicine is an applied science.

The two strands are academic and personal. The first strand deals with my academic career as a science and medical educator, and educational researcher in the

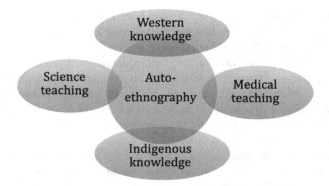

Fig. 1.1 Schematic representation of components of the book

context of different ways of knowing. The second strand concerns my evolving awareness in understanding and appreciating two different ways of knowing—western and non-western, i.e., indigenous African ways of knowing. My intent is to invite readers to share in my journey, to see how difficult it was for me to make the transition from my western way of seeing the world towards gaining an understanding of the African way of knowing—because they are so different.

While these different components and strands may seem incongruent for one book, it turns out that in many respects they are intimately connected through re-occurring themes: Whose knowledge should be taught? What kinds of knowledge are most important? What aspects of the knowledge should be taught? Who should do the teaching? And how should it be taught? The following diagram summarizes the components of this book and the connections between them (Fig. 1.1).

VIDEO: We Can Teach the Children: I interview traditional healers in Lesotho. The Basotho traditional healers explain their ideas about the topics that Basotho children need to learn for healthy living. They suggest how they could teach these topics in conjunction with regular classroom teachers.

We Can Teach the Children: Additional material to this book can be downloaded from http://extras.springer.com

References

Ellis, C., Bochner, A., Denzin, N., Lincoln, Y., Morse, J., Pelias, R., & Richardson, R. (2008). Talking and thinking about qualitative research. *Qualitative Inquiry, 14*, 254.

Hatem, D. (2014). The reflection competency: Using narrative in remediation. In A. Kalet & C. L. Chou (Eds.), *Remediation in medical education: A midcourse correction*. New York: Springer. Chapter 14.

Mullan, F., Ficklen, E., & Rubin, K. (Eds.). (2006). *Narrative matters*. Baltimore: The Johns Hopkins University Press.

Short, N. P., Turner, L., & Grant, A. (2013). *Contemporary British autoethnography*. Rotterdam: Sense Publishers.

Chapter 2
Introduction to Different Ways of Knowing

Abstract The ways of knowing of indigenous Africans and westerners are described. I look at the problem of how westerners discredit the knowledge system of others. I consider the effect of the dominance of the western way of knowing imposed on Africa during the colonial period.

The Lesotho Highlands, 1968

The pilot of the tiny Cessna plane deposited us on a desolate landing strip in the Lesotho mountains. We started our 2-day trek to find an unmapped village known as Seroalankhoane. Our party consisted of Masekhobe who came from this village and served as our guide, David Ambrose, a mathematician dedicated to mapping all of Lesotho, and me, a science educator, both from the University of Lesotho, Botswana and Swaziland.

On the second afternoon of our journey, herd-boys from the surrounding mountain slopes rushed ahead of us to announce the arrival of strangers. David and I were the first Europeans in that place.

The villagers welcomed us with great solemnity. We offered our gifts of food and cooking gear that we brought with us from Maseru. The headman said he would have liked to slaughter a beast for a celebration, but the village was too poor. "Perhaps we would accept a pumpkin?"

We soon learned that Masekhobe's sister was in training to become a traditional healer. She was sequestered in a small hut. Her exposed skin was painted with white clay, and she wore a straw mask over her face so that we could not look at her.

The headman announced that the sister's teacher was visiting that night to instruct his apprentice. The master teacher was a traditional healer from a nearby village.

"You can join us in the celebrations," he told us.

A full moon in the cold winter sky illuminated the mountains around us. We settled into the hut that had been offered us for the night and heated our

(continued)

© Springer Science+Business Media Dordrecht 2015
M.G. Hewson, *Embracing Indigenous Knowledge in Science and Medical Teaching*,
Cultural Studies of Science Education 10, DOI 10.1007/978-94-017-9300-1_2

(continued)

dinner over a primus stove. The nearby drumming was intensifying and I was eager to go and see what was happening. Eventually a messenger announced that we could join the group.

On hands and knees we crept into the dark, smoke-filled hut. A dung fire was burning on the packed earth floor, and cast a red glow over the people crowded into every spot. The master teacher was addressing the group. We had no idea what was being said, but then the dancing started. The apprentice followed her teacher in a call and response sequence. The drummers played faster. The two dancers began to twist and turn until I could see only blurred images. The singing became louder, and more like wailing. The excitement was electrifying, mesmerizing, and ecstatic. The apprentice and her teacher entered a trance state. I could feel the draw of the trance state myself, but I held back and did not let myself slip into it. The trance felt dangerous and out of my control.

We had walked many miles that day. Even though we withdrew, the drumming and singing continued long into the night. The whole event was well beyond my understanding. It was difficult to explain but I knew that I had been privileged to participate in this event. I felt a buzz of excitement that lasted several weeks.

This event stayed with me as a topic I would revisit one day. It was almost 30 years later that I had the opportunity to meet traditional healers (THs) again. I was invited to run a workshop in Mozambique. My workshop was geared to helping masters-level science education students with their projects on alternative conceptions.

In order to model how to do a research project I carried out my own study, explaining the steps of planning and executing a research project. My choice of topic was traditional healers and their indigenous knowledge and skills.

"Aren't you scared of the traditional healers?" they asked me. They obviously thought this was a dangerous project, especially for a westerner.

"No, I am not scared." As a western trained scientist I had no commitment to the African way of seeing and understanding the world. I felt that I was not susceptible to the sort of fear these students experienced around the topic of traditional African medicine and THs.

Indigenous Knowledge (IK)

Indigenous people are located around the world in places such as in Australia, New Zealand, the Americas, Asia, and Africa. Although found in diverse geographic regions, indigenes constitute particular cultural groups, use indigenous languages, adhere to their own political systems, and maintain particular social patterns such as ancestral traditions concerning the seasons of planting, reaping and storing, as well

Fig. 2.1 Lesotho highlands (Photographer: Peter Hewson)

Fig. 2.2 Traditional healer
trainee in Basotho village
(Photographer: Mariana
Hewson)

as birthing and dying, coming of age, and marrying. They identify strongly with their specific geographic areas. Their clothing, preferred foods, singing, and dancing are usually distinctive. Indigenes mainly have a subsistence economy closely tied to their physical environments and weather patterns. In so many ways, indigenes are simply not part of the western industrial world. The biggest difference would seem to be based on philosophical precepts concerning their ways of knowing (epistemology).

Apart from a small book by Vera Bührmann (1984), a Jungian analyst and psychiatrist in South Africa, little has been written about African healing practices. She wrote that the western world is primarily scientific, rational, and ego-centered: 'We are technology and concept oriented—we abstract, analyse, and categorise the external objective world through our thinking and sensation functions.' On the other hand, the African world is 'primarily intuitive, non-rational, and oriented towards the inner world of the collective unconscious' (Bührmann 1984, p. 15). According to Bührmann, traditional Africans 'live closer to the world of the unconscious, where symbols are still alive and vibrant and where archetypal images form a natural part of their daily existence and direct their behaviour in ways that seem irrational to us.... They function largely on the level of intuition and feeling. Images, not concepts, are their main mode of apperception' (Bührmann 1984, p.105).

Indigenous knowledge (IK) is a product of the culture and cognition of people that operate independently of western ideas. African IK predates colonialism and has evolved over time into contemporary forms. One characteristic of African IK is the acceptance that spirit pervades everything (living as well as non-living), leading to a broad concept of spirituality. Traditionally, Africans believe that all spiritual powers come from the African concept of God, and exist in decreasing amounts through various levels such as the ancestors (forefathers), living people, animals, vegetation (plants, crops, trees), rocks and soil, mountains, streams, and the earth itself (Mbiti 1969).

These beliefs lead to the idea that all events in life are caused by the agency of God or the ancestral spirits, and that people have little control over their destiny (determinism); the idea that there is some bigger purpose or eventual end state that justifies the current situation (teleology); non-human phenomena, such as wind or earthquakes can be explained of in terms of human or personal powers (anthropomorphism); and the attribution of animal powers to humans (theriomorphism).

Western Knowledge

Western knowledge is a general word for the dominant knowledge system of the western world that originated in Europe. In many ways, great thinkers in the East (e.g., China, India, Thailand, Japan, Korea), the Middle East, and Africa have contributed to western knowledge. However, modern western knowledge, values, and skills dominate those of indigenous peoples around the world.

The history of western science started with pre-Socratic and Greek notions about the universe. The works of Ptolemy, Copernicus, Aristotle, Newton, Descartes, Darwin, and Einstein, amongst many others, form the basis of western scientific knowledge, values and skills. The philosophical concerns of science include ideas about the nature of reality (ontology), ideas about ways of knowing such as the nature and scope of ideas (epistemology), and the value or worth of different ideas (axiology). In addition, science is concerned with generating ideas on the basis of empirical tests and concrete experiences (empiricism), using reason as the main criterion for believing an idea to be true (rationalism), the idea that everything that happens must have a cause (causality), the chance that something will happen, or not (probability), and the belief that it is possible to explain all natural processes in terms of their functional parts, as if they are machines—particularly relevant in medicine (mechanism and reductionism).

Western Science and African Indigenous Knowledge

In order to describe the different ways of knowing of westerners and non-westerners, I attempt to list some of the characteristics of the respective epistemologies—the grounds of the knowledge systems. Incorporating the work of Aikenhead and Ogawa (2007), and Onwu and Ogunniyi (2006), I list some of these characteristics in the following table.

Some characteristics of African indigenous knowledge and western science[a]

Category	Indigenous knowledge systems	Western science
Knowledge	Necessary for survival, both physical and spiritual	Used for survival, gaining food supplies, health and hygiene, medical advances, national security, economic progress
Nature	Mostly knowable but often mysterious and unknowable. Interdependence of all living and non-living things. All entities are imbued with spirit	Ultimately knowable with ongoing research. Nature is neither good nor bad
Intellectual goals	Seeking harmonious co-existence of knowledge and mystery (especially spiritual forces). Existence of multiple truths	Eradication of mystery by describing and explaining, exploring, discovering, creating, and exploiting for the benefit of humans
Fundamental values	Establishing harmony between humans, nature, and spirits. Respect and equality for all. Participation for the common good. Communalism. *Ubuntu*	Establishing power and dominion over nature for the use of mankind. Knowledge for its own sake. Survival of the fittest. Winner takes all. Individualism
Types of knowledge	Oral knowledge involving appropriate living transmitted inter-generationally in the form of stories, riddles, games, songs, dances, rituals, ceremonies, dreams, and intuitions	Documented knowledge in libraries, texts, publications as well as digitally—involving principles, laws, theories, axioms, conceptual systems, concepts, facts, models, and representations (graphs, pictures, etc.)
Validity	Content/face validity, historic validity, consensus decisions	Content/face validity, predictive validity, concurrent validity, construct validity

(continued)

Category	Indigenous knowledge systems	Western science
General perspective of knowledge	Descriptions of natural and non-natural (spiritual) world: vitalistic, holistic, monist, subjective, relational, place-based, humanistic, deterministic, idiosyncratic, intuitive, spiritual, teleological, and theriomorphic	Explanations and predictions concerning natural world: dualist, objective, reductionist, materialistic, generalized, decontextualized, mechanistic, probabilistic, axiomatic, empirical, rational, and causal

Adapted from Aikenhead and Ogawa (2007), and Onwu and Ogunniyi (2006)
[a]These categories suggest clear differences between the ways of thinking of western and non-western people. They are, however, not completely watertight distinctions—there is overlap and leakage between the concepts in the two columns. Most people think in ways that mix these characteristics, forming a dynamic pattern of thinking that probably changes and evolves during their lives

Process of Making Sense

Sense-making by people in different parts of the world involves different ways of knowing. The western way of understanding the vast array of animals and plants is based on Linnaean taxonomy that reflects either/or methods of categorization. In other words, a biological specimen is classified as either a plant or an animal on the basis of objective morphological criteria, such as the number of petals for plants, or the arrangement of bones for animals. However, many Africans categorize plants in terms of their impact on humans, e.g., being poisonous, nutritious, and/or a healing agent. But these categories are not discrete because there is overlap between them—a plant may be poisonous in some circumstances but a healing agent in other circumstances.

A key difference between African IK and western science concerns the absence of an African commitment to empirical testing and theory building in favor of vitalism (the idea that life is not explainable solely by the laws of physics and chemistry, but also by the presence of a vital life-force), and holism (the idea that the universe and living organisms are whole entities that are more than the sum of elementary particles). When looking at the indigenous medical science of traditional healers, we find approaches that combine empiricism and magic, with physiology, psychology, and sociology (Janzen 1978). These approaches involve the central idea of a supreme being (God) and spiritual forces (recently deceased ancestral spirits) that influence and control the lives and well-being of humans and other living things.

Both medicine and science are evidence-based systems, not religious belief systems. The African way of knowing is not a religion either—no deity is worshiped. Instead it is an ancient and sophisticated (but different) way of seeing and interpreting the world, and explaining the vicissitudes of human lives.

It is easy and commonplace for westerners to discredit other peoples' ways of knowing. For example, Offit (2013) has taken a stand by announcing that there is only one medicine—medicine that works and medicine that doesn't work. He dismisses alternative and complementary medicine, as well as integrative and holistic medicine, and every version of medicine that does not comply with the western way of knowing. He cites scientific illiteracy, denialism, and post-modern thinking as causal agents for the widespread interest in complementary and alternative medicine. And he suggests that because doctors have ceded their medical expertise in favor of joint decision-making (intending to improve patient compliance with medical advice), patients are taking their medical treatments into their own hands.

Medical quackery, charlatanism, and exploitation of desperate patients by unscrupulous healers have been with us for ages. Offit (2013) claims that magical thinking is the underlying reason for much of the public acceptance of alternative healing approaches. He does accept that the western public is dissatisfied with modern medicine, perceiving it to be cold and impersonal. He also concedes that some patients do receive benefits from acupuncture and other healing modalities such as yoga and massage, but states that research must be done to check out the validity of these claims (Offit 2013). Even in contemporary medicine there are 'just some things we can't understand.' For example, why does a virus infect one person, but not another? Is it magic? But Offit claims that while we may not understand such questions at the moment, he is sure that with further research everything will eventually be understood—a type of scientism. This is the view that science is the only reliable source of knowledge, to the exclusion of all other views (Sorrel 1994). According to Blackburn (2005), scientism is a 'pejorative term for the belief that the methods of natural science, or the categories and things recognized in natural science, form the only proper elements in any philosophical or other inquiry.' The attempt to subsume religion to the principles of science is nonsensical. And *vice versa*, to subsume science to the rules of religion would be equally nonsensical.

Traditional healers do not consider their approach as magical. Rather they get their knowledge from their ancestral spirits who speak with them in various ways. The intention of this book is not to judge African indigenous knowledge (IK) as true or untrue, in terms of western criteria. Rather, the intention is to open western minds to other ways of knowing that may be difficult to understand, but which have much to teach us westerners for our own good as teachers and healers.

A caveat to keep in mind is that when describing differences between knowledge systems it is easier to be categorical and insensitive to the shades of grey that exist between the categories. This black-and-white approach is not an accurate reflection of reality—the distinctions between different ways of knowing are fuzzy. But making a categorical distinction is helpful for the sake of explaining differences.

Interactions Between Different Ways of Knowing

Contact between western and indigenous people is increasing in our contemporary world. This is true for the developing countries as well as for developed countries where non-westerners have migrated to the west and joined the school systems. In such heterogeneous situations the 'intercultural rub' can be felt in school classrooms. Western teachers assume they can teach in western ways, but find this is no longer possible. Western medical doctors have patients that come from very different cultures and environments, often bringing their cultural concepts of health and illness with them. Increased cultural interaction provides opportunities for improving many aspects of life, such as in education and medicine. Western science expanded the human capacity for invention, creativity and problem-solving to unimaginable heights. And indigenous ways of knowing produced knowledge and skills that have worked magnificently over thousands of years for the survival of the human species. Westerners can learn a great deal from indigenous people, and *vice versa*. But there are also difficulties when two different ways of knowing intersect.

The western vision of equality in education is considered essential around the world. There are social and economic reasons for a unified western science curriculum. Economic and political influences have resulted in a dominant western curriculum as the norm—every person in every country should be on the same page, the western page.

When people are unable to bridge the cultural gap between western and African societies they are left culturally and morally adrift. And being culturally and morally adrift is dangerous both psychologically and socially.

Four different leaders in southern Africa reflect on the sad fate of a culture that becomes separated from itself as a result of unresolved interaction between different ways of knowing. The first comment is by South African Black Consciousness leader Steven Bantu Biko, writing in the 1960s about the effect of western schooling on African children:

> The African child learns to hate his heritage in his days at school. So negative is the image of African society presented to him that he pretends to find solace in close identification with white society. (Biko, in Stubbs 2009)

During my travels in Lesotho I met Morena Borenahabokhethe Sekonyela, a senior traditional healer who is also a chief, an advocate, and a university lecturer at the University of Lesotho. In our discussion he told me:

> My greatest concern is that western education has produced a gap between what people are, and what they are supposed to be. When people grow up without morals they can turn into monsters, killing with impunity. We need to bring values back into our children who are the future of our society. (2008)

I also met Dr. Lira Molapo while I was collecting data in Lesotho in 2008. He is a Science Advisor to the Ministry of Education, Maseru, and was engaged in research in science education. Lira Molapo told me that in some regions of Lesotho, there was a 0 % pass rate in the O-level science exams.

Science Advisor's View, Lesotho
"My concern about science students is about the environment they live in. It is very much what I would call dilapidated. The land is 'degraded'—there is a lot of soil erosion in Lesotho, there is a lot of pollution, and there are many hazards that are surrounding the Basotho children. And we also have these diseases. They are my concern because there is nothing that the Basotho

(continued)

(continued)

educational system contributes to solving some of these problems—the land degradation, the pollution and the diseases. What is it that the education system should be providing so that the children in Lesotho can survive in a more healthy way? This is my concern.

My research is about linking their formal and their environmental knowledge with their daily practices. My suggestion is that the curriculum should empower our children to take action, to do something about the environment we are living in, to fight the soil erosion we are living in, empower them to do something about the pollution, do something about the littering, do something about their healthy living, eating a properly balanced diet, avoiding the diseases, and living a healthy life. So my suggestion would be to empower our children to take action now, not in the future, even as they are school children. As to the AIDS pandemic they should be empowered to change their behavior. They shouldn't be forced by others to engage in these promiscuous behaviors that lead to the major spread of HIV and AIDS. They should learn they can say 'No!' to whoever comes wanting to manipulate them to have sex. They have to be empowered to take responsibility for their lives… As it is, the children are interested in science but are not passing their exams here in Lesotho."

Interview with Dr. Lira Molapo (June 2008)

Meshach Ogunniyi Ph.D., Senior Professor of Science Education and Director of the School of Science and Mathematics Education, University of the Western Cape, South Africa, told me:

Many Africans throughout Africa have lost track of their cultural identity due to outside influences. Although they try to embrace a new cultural identity they never quite make it. So they are in-between cultures all the time. I find that they tend to denigrate their own cultural knowledge as superstitions, but they don't really understand the western culture except at a very superficial level in terms of material acquisitions. They are neither here nor there. Such people are without roots, have lost their sense of direction, and respect no one. (2010)

Education is a political force—having appropriate knowledge to compete in a rapidly changing scientific modern world is a powerful asset for any nation. But is it necessary to annihilate the IK of a group in order for that group to learn the knowledge of a different, more economically powerful group? I think not. I also do not think it is possible to annihilate the knowledge system of an indigenous group without having a negative psychological impact.

Colonialism and the supremacy of the west have helped create the gap between indigenous people and their current situation. Contemporary education reinforces this gap. Educators must find ways to negotiate this gap. They must be able to reconcile the IK of their students with modern knowledge to enable learning. When teachers fail to take account of their students' diverse cultures, the students often fail to learn—particularly science. And science is the *sine qua non* of modernity, and of the success of countries as participants in world economics.

When seeking to reconcile different ways of knowing, the intention is to heal and rebuild indigenous peoples 'by restoring their ecologies, consciousnesses, and languages, and by creating bridges between indigenous and western knowledge' (Battiste 2000, p. xvii). The ways of knowing of both indigenous and western peoples need to be identified and compared.

These differences are substantial. People representing such different ways of seeing the world may have difficulty communicating with each other. The assumptions of their respective knowledge systems diverge so much that they may be unable to find much, if any common ground (Toulmin 1972). This is especially problematic for westerners when communicating with non-westerners because the western scientific way of knowing has clear rules concerning what counts as valid knowledge, such as rationality and causality.

In heterogeneous communities of people with different ways of knowing, there are numerous occasions when mutual understanding is essential, for example in medical emergencies, in daily trading, and also in classrooms. For functionality within a heterogeneous community, there needs to be interpersonal understanding, dialogue, and mutual respect. Only then can diverse ways of knowing be tolerated and managed by all groups.

My purpose in science education is to find ways of improving science teaching for the greater satisfaction of the teachers and improved examination results for the students. In medicine, I hope to improve both medical teaching and learning (patients as well as medical students) in trans-cultural situations. My aim is to help trainees from non-western cultures know how to manage western knowledge and skills that differ substantively from their own, and for medical personnel such as doctors to better understand their foreign students and care for their foreign patients.

We Can Teach the Children: Additional material to this book can be downloaded from http://extras.springer.com

References

Aikenhead, G. S., & Ogawa, M. (2007). Indigenous knowledge and science revisited. *Cultural Studies of Science Education, 2*, 539–620.

Battiste, M. (Ed.). (2000). *Reclaiming indigenous voice and vision.* Vancouver: University of British Columbia Press.

Blackburn, S. (2005). *The Oxford dictionary of philosophy* (pp. 331–332). Oxford: Oxford University Press.

Bührmann, M. V. (1984). *Living in two worlds.* Cape Town: Human & Rousseau.

Janzen, J. M. (1978). *The quest for therapy: Medical pluralism in Lower Zaire.* Berkeley: University of California Press.

Mbiti, J. S. (1969). *African religions and philosophy* (2nd ed.). London: Heinemann Educational Books.

Offit, P. A. (2013). *Do you believe in magic? The sense and nonsense of alternative medicine.* New York: Harper.

Onwu, G., & Ogunniyi, M. B. (2006). Teachers' knowledge of science and indigenous knowledge and views on the proposed integration of the two knowledge systems in the classroom. In E. Gaigher, L. Goosen, & R. de Villiers (Eds.), *Proceedings of the 14th annual meeting of the Southern African Association for Research in Mathematics, Science and Technology Education* (pp. 128–134).

Sorell, T. (1994). *Scientism: Philosophy and the infatuation with science*. New York: Routledge.

Stubbs, A. (2009). *Steve Biko—I write what I like: A selection of his writings* (pp. 109–135). Picador Africa.

Toulmin, S. (1972). *Human understanding, Vol. 1: The collective use and evolution of concepts*. Princeton: Princeton University Press.

Part II
Science Education

Chapter 3
History of Science Teaching in Southern Africa

Abstract I describe how the Apartheid government instituted the unequal and unfair provision of schooling for black children in South Africa. A brief outline of early educational psychology reflects the distance between, and misunderstanding of western researchers concerning black children. I describe how the increasing stranglehold of Apartheid affected everyone.

Apartheid and Education

In the 1950s the South African government focused more energetically than ever on human engineering to implement its *Apartheid* plan. All Africans were to live in tribal groups, in specific regions set aside for them. The government eventually called these regions 'Homelands,' but they had nothing to do with the normal meaning of homeland. They were repositories for large numbers of people, more like concentration camps where Africans of particular tribal groups who spoke particular languages were concentrated together in geographic regions.

The government forcibly removed large numbers of Africans from their homes to the designated Homelands. After notice was given of impending removal, bulldozers arrived in a village and mowed down the shanty houses. The people protested vigorously but they were loaded on trucks along with their children and aged parents, and their meager possessions. They were then driven to, and dropped off in the designated 'Homeland' regions.

Needless to say, the whites were given the most desirable land from all points of view. It was arable, beautiful, and suitable for industry, with high economic yield. One fifth of the South African population was allowed four fifths of the land.

The Nationalist government also ruled that all students had to attend their own racial universities in various parts of South Africa. This separation was enforced by language: Afrikaans speakers (a derivative of Dutch developed by the first settlers from Holland) went to their own universities such as the University of Stellenbosch, and the University of Pretoria. English speakers went to the University of Cape Town, Rhodes University (in Grahamstown), the University of the Witwatersrand (in Johannesburg), and the University of Natal (in Durban). African students were similarly segregated by geography and language: University of Fort Hare in the

© Springer Science+Business Media Dordrecht 2015
M.G. Hewson, *Embracing Indigenous Knowledge in Science and Medical Teaching*,
Cultural Studies of Science Education 10, DOI 10.1007/978-94-017-9300-1_3

Eastern Cape served Xhosa students, University of the North (known as Turfloop) served the Northern Sotho and Venda students, University of Durban-Westville served Zulus as well as Indians, and the newest, University of the Western Cape (near Cape Town) mainly served the so-called Coloreds of that region.

This chapter considers several of the deeper problems concerning multiculturalism that affect teaching western science in non-western cultures. It suggests the insufficiency of western pedagogy itself.

Apartheid Era 1948–2004

In the period of South African *Apartheid*, Africans suffered enormous inequalities based on racism, a colonial history of subjugation, geographic isolation, and the very real difficulties associated with being schooled in English and Afrikaans when their home language was fundamentally and structurally different. English and Afrikaans are only two of the 11 main languages in South Africa. The poverty/deprivation amongst Africans lasted for generations, and privileged whites would sometimes refer to Africans, as well as the Khoi-San peoples (the indigenous people of the south western Cape Province) as backward. This label implied that nothing could be done to 'civilize' such people.

The idea that the majority of the South African population was backward entered the formal politics of the country. The House of Assembly report in 1953 stated:

> We must accept that the non-European in South Africa, in his level of civilization, is hundreds of years behind the European, and he can only insist on the same privileges and rights as those enjoyed by the Europeans in South Africa today when he reaches that stage that the white man has reached. (Maclennan 1990)

Western educational researchers over the past 100 years focused on how westerners perceive their world, how they develop conceptions about the world, and how these perceptions and conceptions change over time. When applied to science teaching, the questions focused on the ways in which individuals learn science, that is, how they perceive phenomena, how they acquire scientific concepts and how they learn to reason in a scientific manner. Then this research focus was applied to children and adults from non-western cultures.

Much of the early research in cross-cultural psychology involved western experimenters who analyzed the cognitive skills of non-westerners. Wober (1969) suggested that the research question posed by many westerner researchers was: How well can the non-westerner do our western tricks? This question remained at the heart of contemporary discussions concerning indigenous non-westerners and the learning of western science. Wober (1969) criticized this type of research as being unfair and irrelevant.

One example of this type of western-centered research concerned studies on human perception—the processes by which people experience and organize sensory information. For example: How do non-literate peoples and other diverse cultural groups interpret pictorial representations of three-dimensionality? Both Hudson

(1962) in South Africa and Mundy-Castle (1966) in Ghana researched this topic showing that non-westerners were unable to perceive three-dimensionality whereas white secondary school students had no difficulty. The claim for these and similar studies was that western conventions in interpreting pictorial representations were not used by non-western peoples, that is, these skills are learned in western schools and reinforced by western society. They concluded that both formal schooling and western cultural patterns influence the way in which individuals perceive pictorial representations, i.e., that culture and environment affect perception, rather than innate skills (or lack of them).

My Schooling

During my early years on a dairy farm near Johannesburg, I ran about all day in bare feet, playing fantasy games, learning how to milk cows, and offering the new-born calves my fingers to suck. I often experienced the kindness of the African farm laborers who kept an eye on me, finding me when my parents thought I was lost. I started my schooling at a farm school where the children were 'poor whites' from local farms. Everyone spoke Afrikaans. No one, not even the teacher could speak English—and I could not speak a word of Afrikaans. The lessons in Afrikaans mainly passed me by with little impact, except for arithmetic, where numbers transcended language. My parents bought me my first pair of brown sandals and the other children, having no shoes of their own, admired them greatly. At an early age I learned about such disparities. Then my family moved to the city.

The heart of South Africa is the gold mines. Johannesburg, or *Egoli* (The City of Gold) sits high on one of the many rocky ridges that are so typical of this gold-bearing region. The altitude of Johannesburg is about 6,000 feet, approximately a mile above sea level, and you really know it because the altitude makes you feel short of breath and a little light-headed.

We often experienced an underground rumbling that made plates and cups on tables rattle and shake. Underneath Johannesburg there is a maze of tunnels created by miners in their efforts to find and extract the precious metal. The tunnels can be as deep as one mile, where it is so hot that a human body can barely function. These earth movements signified not an earthquake, but an underground rock fall that inevitably meant death for many of the miners. Most of them were groups of African men, each supervised by a white mine captain.

It was 1948. To the surprise of the United Party government under Prime Minister General Jan Smuts, they lost the national election. Under the leadership of Dr. Daniel Malan, the Nationalist Party swept to victory. This was a time when race relations were becoming the central issue. Afrikaner intellectuals had created the seeds of *Apartheid* (separateness) in the 1930s. By the 1948 election the seeds were already sprouting into the doctrine of *Apartheid*:

Each race and nation has its own distinct cultural identity and has been created to fulfill a unique destiny laid down by God. To fulfill its inner potential each nation must be kept pure and allowed to develop freely along its own lines. Excessive contact between the races, above all racial interbreeding, would corrupt and destroy the inner potential of both races involved... On these assumptions, each race and nation in South Africa must have its own separate territory on which to develop along its own unique lines. At the same time, social contact between the races must be reduced to the absolute minimum and sexual relations rigorously prevented. (Omer-Cooper 1987, p. 190)

My family was English-speaking South African. All our forefathers (and mothers) came from the English mid-lands. We tried to speak the King's and Queen's English correctly, and we went to English schools. We drank afternoon tea at 4:00 PM every day, always with a beautiful teapot, hot water jug, cups and saucers, and side plates for the afternoon treats.

At Garden Club meetings, women met to discuss the finer points of persuading roses to survive the African climate. They gathered under the shade of an ancient oak tree, while a bevy of maids set out the tea. Women dressed well for these occasions—their flowery garden hats paid tribute to times gone by. They were rewarded by scones with jam and cream.

In our Africa, despite the rich soil, the wilderness overwhelmed all attempts to create an English country idyll. We softened our African experience with delicate lace and Limoges china tea sets, but we were growing rose gardens where none ought to be.

My Senior School in Johannesburg 1953–1959

I was happy at school. It was a private girls school founded in 1903 in Johannesburg by two women from England. The handsome school buildings had beautifully landscaped gardens. A huge central jacaranda tree graced us with a haze of purple blossoms in spring. The school was founded with a Christian tradition, with the school motto of Honor and Dignity. The school's aim was that the students should study hard, play sport with energy and enthusiasm, obtain academic excellence, and become successful, independent women. We had strong, dedicated teachers and a diverse curriculum.

I loved the school rituals, daily morning assemblies, and the inter-house competitions and celebrations. I also loved the school chapel, a quiet, peaceful place. We often heard the mandate that girls who had the good fortune to attend a school such as ours were duty bound to give back to society. This mantra was drummed into our subconscious minds. The ways in which we might fulfill this mandate of responsibility to the community were seldom discussed.

Somehow our world seemed to focus more on England than on the turbulent political events that erupted in South Africa. School life was so very English, so colonial. We had an unusual school uniform, a sack-like tunic that hearkened back to another of Britain's colonies—India. The desire to colonize Africa was augmented, in part, by schools such as mine.

Empire building by Europeans in Africa dates from approximately the late 1500s. In the quest for tea and spices, first the Portuguese, then the Dutch followed by the English opened sea routes from Europe past the southern tip of Africa to the East Indies. Inevitably South Africa became more than a refreshment station for food and water on this lengthy voyage. In 1652 the Dutch East India Company established their settlement at the southern tip of Africa, in what is now known as Cape Town. French Huguenots who were escaping religious persecution in Europe were invited to move to the Cape region and were assimilated with the Dutch settlers, who ruled the territory in those days. Approximately 150 years after the Dutch settlements, the British took over control of the Cape in 1795, in order to protect their sea route to the east. English personnel elected to stay in what would become Cape Town as part of the team of farmers who provided the water, fresh meat and vegetables for the ships that plied the ferocious seas between Europe and the Indies.

The Dutch and British settlers traded with the indigenous population of the Cape, the Khoi and San people (known collectively as the Khoi-San), descendants of the oldest known people on earth. They were mainly hunters, fishermen, and gatherers. Later, some also became herders of sheep and cattle. Despite the often cruel attempts by the settlers to persuade the Khoi-San to work for them in providing produce for the ships, they refused to do so. Many Khoi-San died as a result of their non-compliance with the settlers' demands. The settlers were desperate for labor. As this was the period of international slave trading, they started importing slaves from the east, particularly from present day Malaysia and Indonesia. They also brought in slaves from Mozambique, Angola, and a small number from West Africa.

In 1884 Africa was carved into different countries by mutual agreement of the colonial powers already invested in the continent. The Portuguese accommodated themselves in Mozambique and Angola; the French in the Congo, Mauritania and Senegal; the British in parts that became Kenya, Uganda, Nigeria, Rhodesia, South Africa and the southern African countries of Botswana, Lesotho and Swaziland. These settlers brought commodities to Africans that Africans grew to like and want—fabric, beads, cooking pots and domestic objects. In return, the traders got ivory, skins, and hunting opportunities. Later on, the discovery of gold, diamonds, and uranium in South Africa, and minerals and precious stones in other regions, powerfully enriched the 'mother' countries.

A remarkable 1912 publication by an author named only as 'Imperialist' provides an upbeat and proud description of British settlers' activities in Uganda. Reading like a shareholders report, the publication documents imports such as private merchandise, government stores and domestic commodities, and exports mainly in the form of cotton, hides and ground-nuts. The author suggests that Uganda has 'many commercial advantages providing an attractive field for investment.' In an exuberant, somewhat romantic tone the author describes the hunting of big game—lion, wildebeest, and elephant, and comments on the opportunities for revenue from shooting parties. The Imperialist also mentions that the East

African Protectorate took care to maintain law and order amongst the native people, commenting:

> As regards the natives everything is done to make them happy and contented, abolition of slavery has been performed in a manner satisfactory to all concerned. Educational advantages have been improved, and a scheme is under consideration for the education of the sons of chiefs, so that in their turn they may instruct their tribes in improved methods of agriculture and sanitation. (Imperialist 1912, p. 79)

Colonialists exploited the resources of Africa for the great economic benefit of their home countries. Successful exploitation needed large numbers of healthy workers who could manage the extraction of essential natural resources. To accomplish this the colonial leaders needed appropriately skilled workforces. The sons of headmen and chiefs were targeted for an education that could fulfill the British policy of indirect rule—a policy whereby London controlled the British colonies. Selected native chiefs, referred to as Native Authorities, were given the responsibility to implement colonial law and education, and collect newly imposed taxes—the hut taxes. These chiefs apparently took a pragmatic view to the advancing colonial powers and invested themselves willingly in the new dispensation as they gained power for themselves, protection from their local enemies, and the ability to acquire the highly desirable goods from European countries. The British colonialists thought of themselves as a civilizing agency in this new world of which they knew very little—a patronizing attitude toward Africa that has persisted to the present day.

The average citizens did not fare as well. For them the colonists provided vocational education in applied topics only. The colonies needed their labor force to be efficient, but not too educated. They required a minimum of basic communication skills and numeracy—a form of education adapted to the so-called needs of the local Africans, or more accurately, the needs of the colonial powers for useful citizens. The new colonial governments needed to generate a supply of somewhat educated and skilled workers, while also denying them civil rights, democracy and freedom, as well as access to material commodities such as land-ownership.

A further dilemma developed concerning African education. On the one hand, some thought that the Africans should not be educated at the expense of their own cultural knowledge. In this respect, culture was seen as the body of knowledge, attitudes and skills in the form of tribal customs, beliefs, practices and moral values.

Long ago, Malinowski (1936) commented that educating Africans out of their culture could be seen as a great anachronism and even a 'serious danger,' creating a rift between the indigenous way of life and the training offered in schools. He stated that a European education would steal the African's knowledge of their own tribal traditions, their moral values and even their practical skills, resulting in a disintegration of society.

On the other hand, Africans needed a modicum of education to be useful in the developing colonies, and to become eligible to hold jobs, receive payment for their work, and increase their ability to obtain greatly desired commodities brought by these foreigners into undeveloped Africa. The African desire to learn the foreign languages was strong. Africans willingly learned whatever they were offered. In Anglophone colonies, an English education was enthusiastically superimposed upon local people.

Traditional educational approaches existed, and included a complex system of formal and informal processes (Hanson 2010). These approaches included children's songs and rhymes, symbolic ritual dances, the telling of stories and riddles, and apprenticeship into various skills and arts such as metal work, weaving, ceramics, beading, building, thatching, planting, reaping, threshing, and animal management. Hanson describes indigenous educational practices where everyone who had learned a skill (such as mother-tongue literacy) was expected to teach this skill to others— a form of 'know one, teach one.' She also described Ugandan educational emphasis on student abilities to generate new knowledge as well as to incorporate their traditional cultural knowledge.

Colonial education, on the other hand, enforced a system of rote learning, especially of number skills, history, geography and science that was derived from, and contextualized within western settings. The various traditional educational approaches were rejected and overwritten with colonial, western approaches and curricula. This policy was similar in all the African colonies.

In South Africa the educational dilemma was resolved by the imposition of the Bantu Education Act, 1953, Law # 47 (Blamires 1955). Under the repressive ideology of *Apartheid*, the nationalist government was not prepared to offer the African people contemporary versions of western education, however intelligent and eager for education they might be. The Bantu, Indian and Colored peoples of South Africa were disallowed the benefits and rewards of having a western education. Under this obnoxious law, the various races of South Africa were relegated to separate educational institutions as part of the *Apartheid* plan of a completely segregated society. The Minister of Education, Hendrik Verwoerd, (who would later become Prime Minister) stated:

> There is no space for him (the 'Native') in the European Community above certain forms of labour. For this reason it is of no avail for him to receive training which has as its aim the absorption of the European Community, where he cannot be absorbed. Until now he has been subjected to a school system, which drew him away from his community and misled him by showing him the greener pastures of European Society where he is not allowed to graze. (As quoted in Kallaway 2002)

During the 1950s the minority government implemented *Apartheid*—the geographic, economic, social, and even religious segregation of its constituent people, with the aim of dominating the vast majority of Africans and other racial groups. In this way, the government could maintain the non-white workforce for the benefit of the whites. The government also reorganized schooling for Africans, making it a second-class educational system, more suitable for educating a workforce of subservient people.

The government systematically closed down mission schools that were giving a select number of Africans a superior education, including future Prime Minister Nelson Mandela and a significant number of anti-Apartheid leaders in the African National Congress. Fundamentally, the *Apartheid* government's aim was to prevent Africans, Indians and Coloreds from receiving the sort of education that would encourage them to aspire to political power, a way of life and economic status that was deemed undesirable. Instead, Africans were to receive an education designed to provide them with skills to serve their own people in the homelands, and to work in jobs under whites in the mines, in industry, on the farms, in businesses, and as domestic servants. Through the Bantu Education Act, Africans as well as other non-whites were provided an education that was neither true to indigenous education nor a reasonable form of contemporary western education. It was a watered-down, inferior, paternalistic education that served to exploit the African people by capping their intellectual opportunities, rather than liberating and developing their intellectual and human potential.

Learning conditions for African students were generally hard, but were made even more difficult in South Africa by the *Apartheid* regime. The Bantu Education Act (1953 #47) increased the gap in learning opportunities between the races: the Whites, Indians, Coloreds and Blacks. Black schools had inferior syllabi, facilities, textbooks and teachers.

I offer a narrative of life in an African school in the 1960s in which I have combined my own experiences of teaching science in a school in Soweto, a black township near Johannesburg, with a recent interview by Wiley et al. (2006) with the Honorable Mr. Obed Bapela, M.P., the African National Congress. Obed Bapala attended school in Alexandra Township, an informal African settlement on the northern outskirts of Johannesburg:

An African School Near Johannesburg, 1960s
The school was an overcrowded place in the middle of Alexandra Township. I used to walk several miles to my school, and I was barefoot. In the winter I would make a kind of protection for my feet by melting candle wax and mixing it with paraffin, and smearing it on my feet like a wax shoe. Our schools had two schedules—there was a morning schedule that started at 8.00 and ended at 11.00. And then those kids went home. The second schedule kids came at 11.00 and went home at 2.00. So the teachers had to run two sets of classes of the same grade each day. Sometimes because of the lack of classrooms we would have our lessons under a tree in the schoolyard. There were between 70 and 80 students in each class. The teachers may have had a textbook but we did not have any textbooks at all.

The teacher wrote important things to remember on the board and we copied them into our exercise books. Sometime the teacher would only state the important facts and we would be like a choir and recite the words back to the teacher, often many times over, until we remembered them. Even though we spoke African languages at home in the township, like Sesotho, or Venda, our subjects were taught in English. History was taught in English and also mathematics and other subjects like science. Science and mathematics had many English words but we did not have the same word in our own languages. We also learned Afrikaans because South Africa was a bilingual country at that time. But what about all our African languages? We did learn some topics in our own languages in the primary school, but later as we became older we

(continued)

(continued)
learned the subjects in either English or Afrikaans. Our teachers carried slim sticks and caned us often. Sometimes the whole class would get caned on our hands. In winter when our hands were so cold it was too painful. One of my teachers called his cane his 'teaching aid.' He would call on a student: "Go and get my teaching aid," and laugh at his joke.
Adapted from Wiley et al. (2006)

During my own schooling during the 1950s and 1960s I had very little idea of the political unrest in the country with the rise of the protest movements by black Africans, and the government's responses to these movements. I confess to a naïveté, of which I am not at all proud.

In 1952, Africans in South Africa mounted the Defiance Campaign. This was an attempt at passive resistance. Small groups of volunteers in the country's major cities protested against the segregation laws by blatantly defying them. For example, they courted arrest by sitting on 'Whites Only' benches. As the campaign grew, it spread to rural areas and the government retaliated. Eventually the passive part of the campaign broke down and rioting started. Altogether some 8,500 people were convicted.

Then during 1956, there were three highly significant events. First, the government disenfranchised about 60,000 so-called 'Colored' people in the Cape Province, by striking them off the voters' roll. Up to that point the Cape Coloreds had enjoyed voting privileges. In August, 20,000 African women marched from Johannesburg to Pretoria to protest the new pass law at the Union Buildings—the seat of government. Finally, in December the government arrested 156 of the leaders of the Congress Alliance and accused them of treason. The Treason Trial lasted for the next five years, and all were acquitted. Meanwhile, the resistance to the government was growing steadily, mainly within the black communities, but a number of whites joined with the black leaders in the battle for civil rights for all, to end the growing oppression in South Africa.

Political Turmoil During the 1950s
Each week my mother visited me at school where I was a boarder. We would sit together in her blue Mini Minor in the school's car park. While we waited for my sister, a day scholar, to arrive we could catch up with each other. She updated me about family news and the political unrest outside the school buildings. This news always seemed as though it came from another planet, far from the beautiful green lawns, the manicured trees and the generally peaceful school environment.

"Did you hear about the bus boycott?" she asked 1 day when I got into her little car. It was winter and I was tired after playing hockey. "I'm planning to

(continued)

(continued)

give lifts to the boycotters this afternoon. It's such a long way for the workers to walk." No, I didn't know about the boycott, and could not figure how so many African workers would ever get home to Soweto or Alexandra Township.

Later she told me that the African women were protesting about the new pass law that made it illegal for them to be found without their reference book. "They're protesting all around the country and hundreds have been arrested." My mother's alarm was palpable. I had visions of the many women that worked as domestics, worrying about the documents that legalized their presence in the city. If a woman crossed the road to visit a neighbor friend, or if she went to the corner café to buy milk, she would be in danger without her documents. This was already the situation for African men, and it was humiliating and enraging for them. I could sense the government tightening its control of Africans' lives.

As a white girl in boarding school, all this news felt like it was wrapped in cotton wool—somehow it did not connect with me on a personal level. Guilt washed over me as I realized how cut off we were in school. But I was proud of my mum for her indignation, and gave her a hug.

On another visit, my mother announced that the government was removing all Africans from another black settlement known as Sophiatown, and bull-dozing their shacks. "They have to go and live in Soweto, but how are they going to do that?" She shook her head, pondering what seemed so impossible. "And Sophiatown is going to become a suburb for whites only." I knew Sophiatown was not so far from where we lived. It was a slum with tin shacks, dusty roads, and few amenities. But it had a vibrant African life with African jazz bands and the popular penny-whistle music that I adored. We had recently hired some penny-whistlers from Sophiatown to play at one of our parties.

"It's not fair, not fair at all." I banged my hands on the dashboard of the car. What was really going on? In this city where I lived I was unforgivably unaware of the ongoing political events. I could hardly fathom the significance of all the bad news.

Next time she visited me my mother told me "There's a treason trial now, and Mandela is accused with many others. But we must not mention his name because he is a banned person—it's dangerous." My mother was grim-faced. She seemed worried and depressed. I knew she was right about the danger. I had recently heard of a man being sent to jail because he drank coffee from a mug sporting a picture of Mandela. All we knew at that time was that Mandela's name was a no-no. We didn't talk about Mandela or the Treason Trial at school. We had little knowledge of the deeper, darker components of the growing anti-government movements. When I returned to school activities, I seldom discussed with my friends these things she had told me. I was aware enough to realize the schism of opinions about racial matters. So I kept the news to myself.

(continued)

(continued)

"By the way, your Uncle David is becoming radical about the South African situation. I fear he may do something rash." Feelings of dread washed over me. Now things were getting closer. David Pratt was my godfather—a special person to me. Apart from his being a successful dairy farmer, I knew little about his political views. This news was ominous. White liberals were being put in jail without charge along with countless Africans, Indians and Coloreds. The killings had started. The car felt hot and claustrophobic—I needed to get out.

"There's been a fire at the Margaret Ballinger Home. Lots of babies and small children died in that fire despite our efforts to call the fire brigade. The firemen didn't come for hours and the damage was done. I think it was intentional." Her eyes welled with tears. "And the government is insisting that we disband the Home and move it into a black area." She was on the executive board of this convalescent home and the news upset her deeply. I squeezed her hand. I too loved the Margaret Ballinger Home, having spent many hours playing with the kids while she attended meetings.

These events troubled me, but I felt far from the action. We were very protected at our school, and apart from my mother's news reports and occasionally reading the *Rand Daily Mail*, there was little discussion about the events happening around us. We didn't know much about the government's efforts to ensure South Africa was under complete control by the whites and, more particularly, the Afrikaans-speaking whites, the Afrikaners.

"Did you know your father thinks *Apartheid* is a reasonable idea? He thinks the means of *Apartheid* could justify the end, and the end is an orderly country that works."

Oh wait a minute! I looked at her in disbelief. Shockwaves went through me. I didn't realize there was a schism in my family. I suddenly emerged out of my sleepy detachment, sensing the very real danger Uncle David was putting himself in, and the unspeakable possibility that my family might be driven apart on political issues. I walked back to my classroom for homework hour, my legs dragging, my world feeling darker than ever before. I was struck by my school's failure to engage its students in this discussion. How was I going to play a role in the increasingly oppressive country of my birth, especially as a boarder at this prestigious school for young ladies?

Following high school, I studied science at the university in Johannesburg and eventually went to England for further study in education. I married and lived in Canada for a while, where I did a post-graduate degree. Then my husband and I returned to South Africa, where I embraced domestic life with the arrival of our two children, cats, a dog, a house in the suburbs, and help from beautiful Magabwe. My personal life was good, very good.

On the other hand, the unrest created by the student uprisings in 1976 resulted in a harsh police response and the deaths of about 600 youth, some of them young children. The school children were protesting about school issues—particularly the fact that they had to learn so many subjects in Afrikaans. The increasing protests concerned Bantu Education and *Apartheid* generally. Children no longer attended school, schools were burned, and the country became increasingly ungovernable.

Post-Apartheid Era

In post-*Apartheid* South Africa, the legacy of earlier, perverted ways of thinking about race has yet to be completely rectified. Twenty years after the end of *Apartheid* in 1994, the educational situation in South Africa is still far from satisfactory. In particular, African students regard science as difficult, with a disproportionately high failure rate in national examinations and the Third International Mathematics and Science Study (Reddy 2006; TIMSS 2003).

The problems facing science teaching in black schools in South Africa are many. Despite the promises of the new post-*Apartheid* government, schools around the country, especially in predominantly rural areas, remain impoverished and unsatisfactory. The antecedents include the geography, history and politics of the region, as well as the various cultures of the diverse indigenous groups. Thus the school problems are not simply due to inadequate financial support, but also to multiculturalism—the problem of identity and accommodation to diversity within a diverse society (Wright et al. 2012).

Many Africans remain impoverished and increasingly discontent with their life situation. What follows is a fictitious interview based on the author's experiences with African students.

Palesa
I am 14 years old and I live with my grandmother in Soweto, near Johannesburg. There are 11 of us kids here, so it is cramped because it is a very small house with only two bedrooms. Some of the kids are my brothers and sisters, and the others are cousins. All our parents are dead from this, um, disease, you know, it's the AIDS. So we live with our grandmother. She has only a little money for all of us and we must help her a lot when we are not at school. She gives us *putu* (corn porridge) and some stew at night, that's it! And she teaches us to pray to our ancestors, especially our parents who have passed. I try and see them in my dreams every night. They help me quite a lot. My school is good and I like to learn. One day I want to be a doctor and heal people who have this terrible sickness AIDS. I like to learn English and numbers. But science! I cannot get it. Science is like magic—you do things and strange things happen, like pink water can change to blue water. I like to learn about flowers and trees, but the names are not the same as my grandmother tells me. I also like to learn about animals, but it is like learning a dictionary with long names and strange drawings. They all seem dead to me, and I cannot remember what the teacher says about them.

Palesa's situation is common in South Africa. HIV/AIDS has ravaged society, leaving myriads of children orphaned or living in dysfunctional households where they care for their sickly parents, look after numerous younger children, and have to find food to sustain themselves and others. The orphans and child care-givers are extremely vulnerable to unimaginable social and health threats such as rape, selling sexual favors in return for food and/or school fees, illnesses such as tuberculosis, sexually transmitted diseases, and of course, HIV/AIDS. Palesa lives with her grandmother, and miraculously, is able to go to school. Despite surmounting all the difficulties to get to school, the system there is far from satisfactory. She usually goes to school on an empty stomach, but finds little to feed her body or her mind, especially in science.

Political, economic and social deprivations are associated with marginalization. People who are marginalized are not catered to in school situations. Even though African children constitute the majority of South Africa's children, and most of them are educationally deprived, the school curriculum targets the minority, relatively affluent children.

The Program of International Student Assessment (PISA) stipulates seven indicator measures concerning adequate educational provision for 15 year olds: a desk for studying, a quiet place to work, a computer for schoolwork; educational software, internet connection, a dictionary, and school textbooks. A child satisfying less than four of these indicator measures would be considered educationally deprived. Palesa has none of them. She is lucky if she eats.

Palesa's complaint about the meaningless drawings in her biology classes might be seen as a consequence of her limited exposure to western ways of seeing and representing the world. If students from different groups perceive pictorial representations according to different conventions, then it is likely that difficulties in interpreting textbook drawings will arise. The use of three-dimensional drawings is common in biology textbooks. In addition, all three sciences (biology, chemistry and physics) use various forms of stylized pictorial presentations such as longitudinal sections, cross sections and magnification at different levels. The research on perception raised doubts about the meaning of textbook illustrations for students whose conventions for interpreting pictorial representations differ from those assumed by the textbook writers. Many, if not most school science texts are written by westerners or authors brought up in the western tradition. Even worse, many science courses offered to students in developing countries originate in a foreign country such as Britain or the United States, and have undergone little modification in the process of export. Many years ago, Dart and Pradham (1967) pointed out that the teaching of science in developing countries was singularly insensitive to the intellectual environment of many students. This is likely to still be the case 40 years after the publication of their research, despite considerable efforts to update curricula and the texts that support them.

The work by the Swiss psychologist Piaget (1972) suggested that the factors that influence children's cognitive development were maturation level, social interactions, and culture. Piaget's descriptions of children's cognitive performance did apply to children in western technological societies. Later on it became increasingly evident that this was not true of Piagetian research in non-western societies (Hewson 1982). There is now strong evidence that schooling affects the development of aspects of logical thinking rather than the presence of universal operational structures as described by Piaget.

In the light of the anomalous findings obtained from cross-cultural studies, Piaget (1972) suggested that the speed of progression through the stages was not the same under all conditions and that retardation of development may be experienced in deprived environments and accelerated in stimulating circumstances. This type of deprivation was thought to result in the inhibition of the normal development of the child resulting in a lower level of intellectual ability, poor memory, and underdeveloped perceptions. Piaget's view contributed to the deprivation theory, which proposed that children from poor homes, slums, over-crowded conditions without privacy and lacking variety in their surroundings tend to be deprived of crucial stimuli for appropriate cognitive development. Such children, it was argued, do not attend pre-schools, are not exposed to books or educational toys, and lack parental guidance and interest. The notion of deprivation was measured in relation to the living conditions of westerners, and by definition most of Africa and the so-called Third World would be classified as educationally deprived.

By the early 1970s the theory of educational deprivation as applied to the influence of culture on cognition became suspect. Cole and Bruner (1971) suggested that methodological problems in cross-cultural testing made the experiments invalid, and that inferences about intellectual incompetence in children from deprived environments were unwarranted.

Several previous researchers such as Keddie (1973), Ginsberg (1972), and Price-Williams (1975) asked why teachers and researchers had failed to see what children from different homes were capable of intellectually. They suggested that assuming that western middle class cognition was the universal norm indicated a bigotry and gross ignorance of other people's ways of life and thought. They also suggested that so-called 'primitive thinking' observed in other cultures might indeed be relatively sophisticated.

The implications of these investigations were that children bring knowledge to the classroom that may be substantially different from that being taught, and that this knowledge reflects the background (culture, environment, religion, experiences, language, and history) of the children rather than their cognitive ability. This research suggested a reassessment of the deprivation theory. Indeed, Wober (1969) suggested that a more pertinent research question should be: How well do people from diverse non-western societies do their own tricks?

Current Situation 2010–2014

In the years after the end of *Apartheid* the income levels of the majority of Africans, which were already appallingly low, actually decreased. But the incomes of a few black South Africans (especially those in government), and white and mixed race

South Africans (particularly those in business) increased by as much as 40 %. According to the 2011 Census in South Africa, black South Africans comprise 79.2 % of the total population of almost 52 million (Statistics South Africa 2013). Nearly 40 % of the African population lives below the poverty line. Violence in South Africa is still high, in the form of murders, rapes, burglaries, muggings, and vigilante beatings.

In recent years there have been protests around South Africa where local citizens have once again created mayhem by stoning cars, and setting car tires alight with petrol, making impenetrable smoke. They have also burned schools, libraries, and other local government buildings. People feel fundamentally unsafe with a police force unequal to the task of controlling violence, and the governmental sectors of education and healthcare suffering accordingly.

The protests focus on the lack of infrastructure in predominantly African townships. Despite the promises of the new African government (led by the African National Congress), there continues to be a lack of adequate housing, electricity and water. African students and black society at large are concerned about the inadequacy and lack of relevance of their education provision (Reddy 2006). Without a strong education system it is difficult for a society to survive and emerge from the inequities imposed on them in the past. The need for substantial and sustained educational reform is urgent. Indeed, there is currently a flurry of curriculum reform in science education that has been turbulent yet exciting.

We Can Teach the Children: Additional material to this book can be downloaded from http://extras.springer.com

References

Blamires, N. (1955). South Africa: The Bantu Education Act, 1953, Law # 47. *International Review of Mission, 44*, 99–101.

Cole, M., & Bruner, J. S. (1971). Cultural differences and inferences about psychological processes. *The American Psychologist, 26*, 867–876.

Dart, F. E., & Pradham, P. L. (1967). Cross-cultural teaching of science. *Science, 155*, 649–656.

Ginsberg, H. (1972). *The myth of the deprived child.* Englewood Cliffs: Prentice-Hall.

Hanson, H. E. (2010). The road not taken: Uganda's village schools. *Comparative Education Review, 54*(2), 1–20.

Hewson, M. G. (1982). *Students' existing knowledge as a factor influencing the acquisition of scientific knowledge.* Unpublished PhD thesis, University of the Witwatersrand, South Africa.

Hudson, W. (1962). Cultural problems in pictorial perceptions. *South African Journal of Science, 58*, 189–195.

Imperialist. (1912). British East Africa and Uganda: Prospects in the protectorates. *The Empire Review, XXIV*(140), 73–83.

Kallaway, P. (Ed.). (2002). *The history of education under Apartheid, 1948–1994: The doors of learning and culture shall be opened.* Cape Town: Pearson.

Keddie, N. (Ed.). (1973). *The myth of cultural deprivation.* Baltimore: Penguin.

Maclennan, B. (1990). *Apartheid: The lighter side.* Plumstead: Tafelberg.

Malinowski, B. (1936). *Native education and culture contact.* Paper presented in Cape Town, SA.

Mundy-Castle, A. C. (1966). Pictorial depth perception in Ghanaian children. *International Journal of Psychology, 1*, 289–300.

Omer-Cooper, J. D. (1987). *History of Southern Africa*. Cape Town/London/Portsmouth: Philip, Currey, and Heinemann.

Piaget, J. (1972). Intellectual evolution from adolescence to adulthood. *Human Development, 15*, 1–12.

Price-Williams, D. R. (1975). *Explorations in cross-cultural psychology*. San Francisco: Chandler and Sharp.

Reddy, V. (2006). *Mathematics and science achievement at South African schools in TIMSS 2003 highlights from TIMSS: The third international mathematics and science study*. http://nces.ed.gov/pubs99/199081

Statistics South Africa. (2013). *South Africa's population*. http://southafrica.info/about/people/population.htm

TIMSS. (2003). *Highlights from TIMSS: The third international mathematics and science study*. http://nces.ed.gov/pubs99/199081

Wiley, D., Pennington, S., & Sleight, J. (2006). *South Africa: Overcoming apartheid, building democracy*. Part of the Matrix project, Michigan State University. Retrieved from: http://overcomingApartheid.msu.edu/video.php?id=40

Wober, M. (1969). Distinguishing centri-cultural from cross-cultural tests and research. *Perceptual and Motor Skills, 28*, 488.

Wright, H. K., Singh, M., & Race, R. (2012). *Precarious international multicultural education: Hegemony, dissent and rising alternatives*. Rotterdam: Sense Publishers.

Chapter 4
Teaching Science in Southern Africa

Abstract In the context of early experiences of teaching in South Africa and Lesotho, I describe how my curiosity led to two research projects concerning density and heat. I focus on theories of teaching and learning, especially students' acquisition of scientific knowledge. I discuss the role of alternative conceptions in learning, and the implications for curriculum.

It was 1960 and I was passionate about the natural sciences, particularly zoology and botany. It seemed obvious that I should follow my dream to study science. I also wanted education classes, as I assumed I would eventually teach.

Just before my first term at university started in February 1960, Britain's Prime Minister Harold Macmillan delivered his Winds of Change speech to the South African parliament. He warned of rising African consciousness throughout the African continent. He also warned that South Africa's policies of *Apartheid* were anachronistic to the world. It was now time for the South African government to start providing rights and opportunities to Africans.

With Harold Macmillan's warning words ringing in my ears, I enrolled at the University of the Witwatersrand (the ridge where the waters are white) in Johannesburg. The university excelled in mining engineering and medical sciences. It was a bastion of liberal thinking during the 1960s. While I attended the university, ongoing student protests for the restoration of academic freedom for all South Africans dominated my life.

An imposing set of buildings housed the academic departments. The Central Block housed the university administration and had a flight of steps that led up to Grecian columns and the front door. This was where students met during the break periods—the social center of the university. There were manicured lawns with large chestnut trees near the library. The student buzz was intoxicating to me.

First Teaching in Johannesburg
I was studying science and I also felt the urge to try my skills as a teacher. So I volunteered to teach biology in the African Night School organized by the university.

(continued)

© Springer Science+Business Media Dordrecht 2015
M.G. Hewson, *Embracing Indigenous Knowledge in Science and Medical Teaching*,
Cultural Studies of Science Education 10, DOI 10.1007/978-94-017-9300-1_4

(continued)

After my afternoon zoology lab I walked over to the Central Block to find my students. I felt quite confident and eager for my first teaching experience. After all, I loved my studies in botany and zoology, I had always wanted to teach, and I felt a great empathy for the plight of African students, many of whom had been unable to get to school.

These men and women were now eager get their school completion certificates. All of them had already put in a full day's work on the campus, cleaning offices and classrooms, cooking in the cafeteria, and doing all the jobs that keep a big urban university running smoothly.

They had never had the opportunity to sit as students in the university classrooms. Instead of entering by the back entrances of the main building, they had climbed the many steps leading up to ionic columns and had entered through the glass, double doors of the university. The corridors were wide and sonorous, encouraging tentative whispers. The classroom had a high ceiling and wood paneling. The pictures on the walls were of venerable university teachers—white men in dark suits. They created a very different picture from myself—a young white inexperienced female teacher in a lab coat, and my students—black men and women from the various townships surrounding Johannesburg.

My students looked both eager and anxious. They sat silently in their chairs awaiting their teacher—me. I suddenly wondered how I had the audacity to take on this task. But I was committed and there was no escape.

I had prepared a lesson on The Plant. This should be a nice starting point, I thought. Every one of them would be somewhat familiar with plants, especially as some of my students were gardeners.

At my first class, I greeted about 25 adult students, men and women in English. They responded in chorus, "Good afternoon Ma'am."

I carefully introduced myself, and the topic of instruction. Then I drew a picture of a plant on the board, labeling the roots, stems, leaves, and flowers.

"What is this that I am drawing?" I asked questions sporadically in an attempt to engage them. But no one answered any of my questions. Engagement was nearly impossible. I felt as if I was dragging myself through a swamp, without being able to get purchase on firm ground. The experience somewhat deflated my optimism. I was surprised at how challenging this teaching project was turning out to be.

It will be better next time, I told myself. At least they will recognize me, and I will know some of their names too.

The second lesson started with me reminding them of a few simple points from the first lesson.

"You remember that a plant has roots, a stem, leaves, and flowers?" I asked hopefully.

(continued)

(continued)

Silence. I looked from face to face, seeking a sign of acknowledgment, a show of engagement, perhaps a flicker of a smile. But there was none.

I tried again: "What did we talk about last week?"

Silence. Passive, blank faces looked back at me. I shifted uncomfortably, silently imploring the students to respond.

Feeling a bit desperate I asked: "Can you tell me one part of a plant?"

"*Ha re tsebe* (we don't know), Ma'am," volunteered a man in Sesotho. The group nodded—they agreed, they did not know.

I was dumbfounded, and continued with a lecture about stems, even though I had no idea what they heard me say, how they interpreted my drawings, or whether they understood anything. Perhaps they were unable to speak English. To be honest I did not even know what levels of education they had achieved, or when they had last been in school. It was probably years ago for most of them. It was also likely that some had had very little school experience. Feeling like a failure, I dropped out of the program and went back to focusing on my own studies, which were in dire need of attention.

Years later, I learned that for traditional Africans it was considered rude to display their knowledge in front of the teacher, and the most polite approach was to announce complete ignorance on the topic of instruction. Furthermore, it was not customary to discuss matters with teachers. Rather, students simply listened and tried to memorize the content presented to them.

Until this point, I had assumed that the gap between Africans and the whites of South Africa was simply economic, a matter of deprivation. I had no excuse for this naiveté, as I lived with Africans in my parents' household. Yet this teaching event was my first conscious awareness of a cultural gap that probably affected the effectiveness of teaching approaches that I took for granted as well as many other facets of life, such as providing medical care.

Learning Science

In general, learning science requires a change from existing knowledge (knowledge accumulated from previous experiences) to orthodox scientific knowledge (formal knowledge learned in school). School learning difficulties experienced by non-western students may be compounded by factors such as their non-western cultural and religious beliefs, as well as their history and physical environments, and the fact that they often learn science in their second language.

During the colonization of Africa, western worldviews accompanied the western settlers. Western science constitutes a cultural hegemony—the result of the dominance of a western philosophy over the indigenous people of Africa. This cognitive domination of a non-western society by invaders is mirrored in the classroom where the worldview of the dominating culture becomes powerful. In some colonized countries this process has been more violent—where children have been forcibly removed from their homes and sent to western types of boarding schools (e.g., Australia, Canada, parts of Africa). Other types of worldview domination have been more insidious, where schools serve as a medium of domination.

Conceptions are products of the mind and are functional units of thought. They are functional and viable ideas that exist within these knowledge systems—they serve to predict and explain the world in which a person lives. Different conceptions can be found in the different thought systems of people originating in diverse cultures. Toulmin (1972) proposed that knowledge develops as a result of the dynamic interaction of each person with his/her environment, especially the intellectual environment. He coined the term 'conceptual ecology' to describe this interaction. Each person adapts to the intellectual niche in which he/she lives. This adaptation causes each person, or group of people to develop knowledge systems (or worldviews) in different intellectual environments that may differ fundamentally from each other. These can be termed 'alternative conceptions.'

Alternative conceptions can be different from, and sometimes at variance with orthodox, western scientific conceptions. Alternative conceptions have been identified within western societies where students in schools appear to have strongly held ideas about the world that differ fundamentally from those of western science. Many science education researchers around the world documented the existence of alternative conceptions in widely diverse science topics. Rosalind Driver (1981) was one of the early researchers who established the presence of alternative conceptions within the cognitive frameworks of science students in England. Her work contributed significantly to constructivist learning theory, and spawned the proliferation of similar research internationally, including within western cultures (Duit 2002). There is now extensive evidence of the existence of alternative conceptions in children's understanding of natural phenomena. And these alternative conceptions interfere with learning science, even within western classrooms.

Unfortunately, when westerners notice the different knowledge systems of indigenous people (indigenes), they tend to make pejorative judgments about the mental capacity of 'the other.' In South Africa, Africans have been labeled as backward and unfit for learning science and mathematics as well. However, Lévi-Strauss (1966) and Horton (1967) described the conceptual systems of non-western peoples as involving kinds of science that are valid in that they serve to make sense of the real world and are the consequence of individuals' interactions with the world in which they live. Although the conceptions of non-westerners may differ from those of westerners, the commitment with which they interpret and explain their world is essentially the same.

From the western perspective, indigenous knowledge in Africa can be, and has been labeled as involving alternative conceptions. In both, the conceptions are alternative to those of western scientific knowledge. However the indigenous knowledge of other cultures is deeply rooted in environmental, historical, philosophical, and social contexts, whereas the alternative conceptions reported in western school children may be idiosyncratic conceptual constructions to explain worldly phenomena as they experience them. In both cases, these ideas are strong and resistant to change. More recently, western theoretical ideas concerning alternative frameworks in other cultures become less hegemonic. The theoretical ideas have changed in order to be more accepting and respectful of the validity and importance of the knowledge of non-westerners, especially indigenes.

African Indigenous Knowledge

Indigenous knowledge (IK) is a product of the cognition of people that predate westerners, and have operated independently of the western paradigm for thousands of years. African IK predates colonialism and has evolved over time into contemporary forms of IK. Even within western societies there is evidence of IK, especially in heterogeneous communities.

IK concerns all aspects of life such as building, planting crops, cooking, baking and brewing, formal social relationships and events (birth of a baby, marriage, sexual behaviors, dying, and death), legal concerns, economic systems, and medicine. IK that is scientifically relevant includes knowledge of survival strategies concerning agriculture, fishing, forest resource management, astronomy, climatology, architecture, engineering, medicine, nursing, veterinary science, and pharmacology (Odora Hoppers 2002). IK also includes indigenous approaches such as teaching strategies, school and educational systems, systems of medical practice, as well as different research methods (Bishop 2011).

One characteristic of African IK is the acceptance that 'spirit' pervades all things (living as well as non-living), leading to a broad concept of spirituality. Traditionally, Africans believe that all spiritual powers come from the African concept of God, and exist in decreasing amounts through various levels such as the ancestors (especially the 'recently departed' relatives), living people, animals, vegetation (plants, crops, trees), rocks and soil, mountains, streams, and the earth itself (Mbiti 1969). While conceptions of 'spirit' are not relevant in western science, they are extremely relevant within the African worldview and cannot be overlooked.

The articulation and documentation of the IK of indigenous societies has become important in numerous parts of the world, such as Australia, New Zealand, South Africa, Brazil, and Canada. Moreover, researchers are working to understand the epistemologies of different indigenous knowledge systems (Barnhardt and Kawagley 2005; Aikenhead and Ogawa 2007). These efforts are, however, at an early stage. The African IK system remains obscure and even unknown to many people who grew up in western environments.

Curiosity

My early teaching experience in Johannesburg and later on at the University of Botswana, Lesotho and Swaziland planted a seed of curiosity about different ways of seeing the world by westerners and non-westerners. I knew about African culture and tribal customs, but the implications of different mindsets and worldviews intrigued me. Even though I grew up in South Africa, it had never occurred to me that there was a gulf between the African ways of knowing and my own.

Teaching in Lesotho – Late 1960s

It was a warm summer day in Lesotho, the kind where the heat shimmers and the bees buzz about their business. Blue mountains encircled the university, a European-styled institution that still served three African countries: Botswana, Lesotho, and Swaziland. To my great joy, I was a temporary lecturer at this university, where I was immersed in the challenges of teaching science education to potential African teachers.

It seemed like a good idea to organize a Science Fair for school children in standards 6, 7, and 8 (middle school equivalent). This would allow me some insight into the mindsets of the kids and their teachers. I wrote the instructions, set the date, and busied myself getting the word out to the schools. Having never organized a Science Fair before, and not having a clear idea of which school teachers had responded to my invitation to participate, I was unsure of what was going to happen. While I was anxious about the mechanics of this Science Fair, I was mostly curious about how the teachers would respond to the challenge, and what sort of questions the kids would tackle. I knew that science subjects were difficult for students in Africa and not perceived as particularly interesting. Something about school science doesn't ring right, doesn't sound true, and doesn't automatically make sense to them. The truth is, I had no idea what might turn up. Trickles of sweat ran down the back of my legs and my back.

School kids from local schools around Maseru and Roma would soon be coming to the zoology laboratory at the University of Botswana, Lesotho and Swaziland. The empty lab smelled of death and formalin. Old zoological specimens lined the walls. Samples of fish, frogs, snakes, mammals and even the mandatory human fetus were stuffed into their glass jars, inert and unlovely. Insects were pinned on boards and mounted behind glass frames on the walls. At least the butterflies maintained some of their original beauty. I wondered what the African kids would think of all these dead creatures preserved in jars and on boards.

A handful of teachers with their students showed up. I welcomed the group, biting my lip that the numbers were so small. They arranged themselves

(continued)

(continued)

around the laboratory, paying little heed to the dead specimens. Two girls set up their exhibit on a bench near me. They wore their black and white school uniforms with hair neatly braided into geometric designs.

"Come, come, Ma'am. Come and see ours first." Huge smiles accompanied their halting English. They were clearly excited.

They showed me two baking trays filled with soil. One had green sprouting bean shoots. In the other, the bean shoots were withered or dead.

"OK, tell me about your experiment," I invited them. "What did you do?"

"We planted bean seeds in here and also here. This one we gave some water, but not this one," volunteered one of the girls.

Sounds reasonable, I thought. "So what is your conclusion?"

"Ma'am, we can say that if you love the plants they will grow. If you hate them they will die."

Now wait a moment. I had assumed that the intervening variable was the presence or absence of water. And now we are talking about love? I looked at them silently, trying to hide my surprise. In all my studies of science where did the issue of love ever feature?

"Tell me more," I urged, hoping to understand their line of thinking.

"Ma'am, we now know that if you love your plants you will give them water, and they will grow nicely. But if you hate them, you can refuse to give them water. See, they have died."

I recalled my prior experiments at the university in England where I was studying circadian rhythms. To do this I had analyzed locust exoskeletons—slices of locust leg, to be exact. This research involved the deaths of numerous locusts. In the locust house on the rooftop, I first ensured the locust eggs hatched. Then I fed the little critters with sweet tender grass. They are cute at this stage, like Disney versions of Jiminy Cricket. When old enough I submitted them to awful deaths in order to get the necessary tissue samples. This eventually made me sick at heart. Even now, my skin crawls and my stomach cramps as I remember the endless cycles of locust growth and death, and the nauseating smell of adult locust dust. The locust holocaust got to me.

Yet, as a scientist, I was doing exactly what was expected of me. I was being objective and dispassionate. The means of the experiment justified the potential end results. I didn't hate the locusts any more than the students hated the plants that they did not water. But both the students and I were interfering with the naturally beneficial conditions that normally support life. In this science lab I was meeting an intersection between two world-views: one in which scientific experiments are carried out with objective dispassion, and the other in which love promotes growth and hate causes death.

Suddenly I appreciated the African view that illness and death can be caused by malevolence—people with anger, fear, resentment and jealousy. And wellness is the result of personal equilibrium, interpersonal harmony and love.

It was time for study – again. I signed up for my doctoral studies at the University of the Witwatersrand in Johannesburg. My undergraduate and master's level studies in science and education theory led me to research more optimal ways to teach science in non-western communities. I decided to analyze African students' conceptions of world, especially concerning basic physics.

The South African Department of Education and Training permitted me to collect data for my doctoral research in a remote part of South Africa known as Qwa Qwa. Those were the days of *Apartheid* and the Nationalist government had established Qwa Qwa as one of the homelands for South Africans of Basotho descent. Located in the rain shadow region of the Drakensberg mountains, the place was both breathtakingly beautiful and extremely cold in winter. In the distance, the easterly range of the Drakensberg mountain tops edged the sky with lace. Winter snow on these mountains caused a sharp wind to cut through the land and its people. Nothing grew in this region, the soil was too shallow and parched. The inhabitants were mainly old people who lived off scanty pensions. Children populated several primary and secondary schools. Their performance in the national examinations was dismal.

Small children usually gathered along the roadsides, playing with wheel rims of old bicycles that they rolled with a stick. Many of the younger boys played with toy cars they made from wire, with moveable wheels and steering devices. They were mostly showing off their handiwork to each other.

These children are bright, I thought. It takes great skill, precision, and imagination to create these wire toys. I wondered at the mathematics, geometry and physics embedded in these creations. Perhaps I could research the knowledge and skills of the preschool kids through analyzing their handmade toys. But I already had a commitment to a different project—testing and analyzing the knowledge and skills of high school students concerning the concepts of mass, volume and density.

How Do Students Acquire Scientific Knowledge?

I became interested in the teaching of orthodox western science to students in a non-western culture. I wanted to understand particular students' cognition in the context of their conceptual ecology (Toulmin 1972). If I could find out more about the indigenous knowledge and skills of indigenous children, I might learn some powerful things. This would be a way of answering Wober's (1969) question: How good are the non-westerners at their own tricks?

In the tradition of cross-cultural research in the 1970s my study involved subjects from a non-western cultural setting as an opportunity to clarify theoretically important variables. The students were from the Basotho linguistic group, whose living conditions were non-technological, and whose home language was Sesotho. They had been exposed to orthodox western conceptions in science in their high schools.

I was interested in identifying, analyzing and then representing the students' knowledge structures concerning the floating and sinking of objects when put in water. It seemed that floating and sinking would be common phenomena in the experience of everyone. And high school students would have been taught the basic scientific concepts of mass, volume, and density at some point.

Forty high school students (including boys and girls) with an average age of about 18 years were selected from the Form 3 class (10th grade equivalent) from four high schools in the Qwa Qwa area in South Africa. The initial selection requirement was that the students should have better than average scores in English, as the interviews were in English.

Basotho children have a home language, Sesotho. In the 1970s they normally began learning English and Afrikaans in the primary grades, where the medium of instruction was their home language. Then in high school they learned science in the medium of English.

While the schools visited for this study were among the best in the area, they were relatively poor. Their limited financial resources meant that facilities such as school laboratories and libraries were poorly equipped. The teachers were well motivated but were often inadequately trained in their subject areas, particularly in science.

I was interested in the students' conceptions of mass, volume, and density. In order to establish the nature and extent of each student's knowledge about the phenomena of floating and sinking I replicated a task from Inhelder and Piaget (1958, p. 20). Each student was first asked to sort the various objects into piles of those that would sink and those that would float, and offer a reason for his/her decisions, and then invited to experiment with the objects. During this process, the student was asked nondirective questions to encourage him/her to enlarge on the statements made or the actions taken. The interviews were recorded and transcribed.

I analyzed the responses of the students for the essential ideas. These statements were summarized and compiled into a Conceptual Profile Inventory (CPI). Selected individual students' statements were used as exemplars of the main CPI categories.

The CPI categories involved both scientific and non-scientific conceptions of mass, volume, and density. I identified these ideas as alternative conceptions. These included conceptions that: mass is a property of only some objects but not others, e.g., a brick, but not a feather; mass is relative (not absolute); mass equals density, and also mass equals heaviness. In the students' view, volume is the size or quantity of something. It has the same meaning as capacity, so some objects have volume (e.g., water in a cup) while others do not (e.g., a pin). A change in the shape of shape of an object, moreover, results in a change of its volume. Concerning density, the students believed that density is the same as mass. They also explained density in terms of denseness or crowdedness: density concerns the packing of particles—closely packed particles increase denseness but loosely packed particles are not as dense.

Many students fluctuated between using both scientific and alternative conceptions. Most of the students attempted to look for causal reasons to explain the floating and sinking of objects in water. One of the students offered idiosyncratic reasons i.e., each object had a different reason to account for floating and sinking based on concrete observations. Only one student realized the clumsiness of his line of argument and resorted to the more parsimonious description of scientific density as mass per unit volume.

In general, I noted that important scientific concepts were missing, e.g., the conception of volume and the conception of particles having mass. In addition, alternative conceptions concerning mass, volume, and density were present.

The conceptions used by the students in the floating and sinking task can be interpreted in the context of their intellectual environment and home language. For instance, the finding that the scientific conception of volume was all but absent could be explained by the apparent lack of usefulness of the conception in these peoples' everyday language (there is no equivalent word in Sesotho). The alternative conception of density can be explained in terms of the everyday use of the Sesotho word, which is similar to the English everyday word 'dense.'

Apparently the scientific conceptions introduced in the science classroom had not replaced the students' alternative conceptions concerning mass, volume and density. These data showed that in some cases both scientific and alternative conceptions existed together as an unresolved anomaly (Hewson 1982, 1986). In a recent paper Ogunniyi (2004) called this situation 'equipollence,' where two competing ideas in a person's cognitive framework have equal cognitive power and co-exist without logical connection between them.

What are the implications of these findings for science teaching within western cultures? How should school curricula be designed? What instructional strategies could be employed? Furthermore, what are the implications for interactions by westerners with non-westerners outside of the classroom, e.g., how do physicians communicate with and teach their foreign patients?

At this time (1970s and 1980s) research on alternative conceptions blossomed around the world both in western and 'developing' countries. Such results stimulated tremendous energy in discerning the role of subject matter knowledge in learning—what was being learned was just as important as the psychological processes of learning. Constructivism took root and flourished.

Tracking an Alternative Conception of Heat

About this time I was employed as a researcher at the National Institute for Personnel Research, in Johannesburg, South Africa. I had stumbled on several anthropological studies about a powerful idea concerning heat used by Africans living in the hot, dry northern part of South Africa, e.g., Hammond-Tooke (1981). I remembered an event that I had never been able to explain.

Being Hot in Johannesburg, 1973

"Let's go out to dinner." My husband went to negotiate with our baby-sitter.

Magabwe was from Venda, one of the homelands in the northern part of South Africa. She came to Johannesburg, the 'City of Gold' to work for us as a nanny. Being a rural person she was traditional in her ways, and her English was limited.

Magabwe arrived full of smiles and took over the business of getting our children into bed. We made our escape.

"How were the children?" I asked when Magabwe met us at the door. I was not expecting any ominous news.

(continued)

(continued)

"The baby was hot," she answered in a matter-of-fact, unconcerned kind of way.

"And now?" I asked.

"No, she is fine now."

While my husband was paying for the baby-sitting I got the thermometer from the cabinet, and put it in the sleeping baby's armpit. Her temperature was absolutely normal.

Mystified, I checked again. "How hot was she, Magabwe?"

"Not too much hot," Magabwe's demeanor was completely calm, as always.

I checked the baby's temperature once again. Normal. Still mystified we decided to call it a day and see what the morning would bring. The next day her temperature was still perfectly normal.

Several years later I learned that many traditional southern African people have a particular form of indigenous knowledge that attributes the word 'hot' to a human state of disequilibrium. Magabwe's use of the word signified that the baby had been fretful, crying, or unhappy. It had nothing to do with having a fever.

The 'Heat' Study

Following the incident with Magabwe and the baby, I embarked on a study to investigate a logical fit between the intellectual environment of northern South Africans and their conceptions of both metaphoric and physical heat. My long-term interest concerned how these and other similar ideas might affect students' learning of science in school as well as in medical practice.

My colleague Daryl Hamlyn and I set out to find empirical evidence for the contemporary existence of the cultural metaphor concerning heat (Hewson and Hamlyn 1985). The cultural concept of heat is a type of metaphor. It concerns a person being ill at ease, such as when ill, pregnant, traveling great distances, and when a person has died. Traditionally, the cultural concept of heat has little connection with the physical concept of heat.

We interviewed 20 people, all of whom represented the Northern Basotho group, who were living in urban or semi-urban areas, and whose home language was Northern Sotho or Setswana (10 high school students, four unschooled adult workers, and six semi-schooled adult workers). There were equal numbers of males and females.

We used photographs to represent life situations involving pain, fear, or distress (sick children, man returning home, lightning, woman birthing a baby, etc.).

We interviewed each person and invited him/her to discern whether (or not) each situation involved 'being hot,' and to offer reasons for their statements.

Our interview results showed that 80 % of the respondents identified the metaphorical conception of being 'hot' in several life situations. This showed that the heat metaphor was still in use by Sothos who were urban and somewhat acculturated to western living. For example: people who are tired/exhausted from travelling long distances; people who become impatient or restless, for example from waiting a long time; women who have recently given birth, are pregnant, or are menstruating; an angry person; many forms of illness involving pain (not to be confused with fever); a dead person and the bereaved family due to feelings of grief; persons killed by lightning; and boys involved with initiation (undergoing demanding puberty rites and challenges). Moreover, people who are 'hot' due to any of these situations are thought to be in an unnatural state in which their blood 'is fighting with itself' causing agitation, distress, or being unsettled.

Then we provided simple physics demonstrations involving heat (boiling water, melting wax, expansion and contraction of metal, conduction of heat in a metal rod, etc.). We asked the respondents to explain what was happening. Some of the respondents offered a pre-kinetic version of the nature of heat. Most of the respondents (85 %) revealed several alternative conceptions concerning heat, for example: observable physical phenomena (smoke, steam) cause the expansion of an object; the particles of a heated object multiply or split to increase the number of particles, thus occupying more space; the particles themselves can expand when heated, thus occupying more space; heat is something that comes into the object, making it bigger; and the packing of the particles of an object influences the way heat is conducted. The cultural concepts concerning 'being hot' can be seen as a metaphor. This metaphor promotes the idea that coolness is good (as in social harmony, personal contentment and well-being), and the converse idea that heat is bad (as in social disharmony, personal distress and ill-health). In the South African context, the hot dry climate that is prone to drought and great hardship is the metaphor for a negative state of being, while coolness is the metaphor for well-being. Metaphors are not simply linguistic structures, but they influence the way we perceive and conceptualize phenomena, and also how we act. People who use this cultural metaphor in everyday life may well use it to try and understand new concepts in science.

Alternative Conceptions

These two research studies concerning heat, and mass, volume and density, raised theoretical questions concerning the role of language and metaphors in an intellectual environment. We were now curious about their contribution to the existence of alternative conceptions, and their place in the conceptual ecology of a people. Within a conceptual ecology there are conceptions that may differ from those of the western intellectual tradition. In this context, it seems commonsense that a people's culture, socio-economic status, history, religious beliefs, myths, local geographic situations, and languages influence how they see their world.

To use the language of conceptual change (Posner et al. 1982), alternative conceptions are intelligible, plausible and fruitful to the person who uses them. When alternative conceptions come into conflict with the ideas taught in science classrooms, there may be serious implications in terms of what the student can, or will learn (Hewson and Hewson 2003).

Implications for South African Science Curriculum

In 1994 Nelson Mandela became President of South Africa and prepared to normalize the country by redressing the atrocities of the *Apartheid* regime. The new African National Congress government began a program of restructuring education and introducing new, and improved curricula.

The new education system introduced a revised curriculum in 1997 based on outcomes-based education policy. It was further revised several times. Of significance, the new curricula specified that 30 % of the science and mathematics curricula should incorporate some of the indigenous knowledge of the vast majority of the children in school, the Africans. In other words, science and mathematics should be applicable to real-life situations and incorporate the cultures, beliefs, and worldviews (indigenous knowledge) of the diverse cultural groups that make up South Africa.

Unfortunately the curricula did not offer suggestions for how this could be accomplished. It is currently an ambitious and enlightened curriculum, but not without its critics. Many teachers felt they had not been consulted about the new topics, especially concerning indigenous knowledge. And, moreover, did not know much about indigenous knowledge nor how to teach it. There have been subsequent modifications to this curriculum in the last 10 years.

Despite these innovations, it can be no surprise that the international comparative study of scholastic achievements (TIMSS 2003) showed South Africa at the bottom of the achievement list for Grades 8 and 12. Other countries in Africa, such as Lesotho, perform poorly too, especially when compared with many Asian and European countries.

The tension that underwrites the teaching of western science in non-western communities such as in southern Africa has come to a critical point in current educational research and practice. These tensions are reflected throughout the western world, especially in classrooms of contemporary American and European schools. These classrooms have become microcosms of the multicultural world. No longer do we need to travel to far-flung remote corners of the earth to find people with ideas at variance to those that dominate in the west. These ideas are within current classrooms, whether or not they are recognized or admitted into classroom dialogues. Inattention to the nature of these knowledge systems can lead to failed communication and under-achieving students.

We Can Teach the Children: Additional material to this book can be downloaded from http://extras.springer.com

References

Aikenhead, G. S., & Ogawa, M. (2007). Indigenous knowledge and science revisited. *Cultural Studies of Science Education, 2*, 539–620.

Barnhardt, R., & Kawagley, A. O. (2005). Indigenous knowledge systems and Alaska Native ways of knowing. *Anthropology & Education Quarterly, 36*, 8–23.

Bishop, R. (2011). *Freeing ourselves*. Rotterdam: Sense Publications.

Driver, R. (1981). Pupils' alternative frameworks in science. *European Journal of Science Education, 3*(1), 93–101.

Duit R. (2002). Bibliography – *Students' and teachers' conceptions and science education*. http://www.ipn.uni-kiel.de/aktuell/stcse/bibint.html – sdfootnote1sym.

Hammond-Tooke, W. D. (1981). *Patrolling the herms: Social structure, cosmology, and pollution concepts in southern Africa* (Raymond Dart Lecture No. 18). Johannesburg: University of the Witwatersrand Press.

Hewson, M. G. (1982). *Students' existing knowledge as a factor influencing the acquisition of scientific knowledge*. Unpublished Ph.D. thesis, University of the Witwatersrand.

Hewson, M. G. (1986). The acquisition of scientific knowledge: Analysis and representation of student conceptions concerning density. *Science Education, 70*(2), 159–170.

Hewson, M. G., & Hamlyn, D. (1985). Cultural metaphors: Some implications for science education. *Anthropology & Education Quarterly, 16*, 31–46.

Hewson, M. G., & Hewson, P. W. (2003). The effect of instruction using students' prior knowledge and conceptual change strategies on science learning. *Journal of Research in Science Teaching, 20*(2), 383–396.

Horton, R. (1967). African traditional thought and western science. *Africa, 37*, 50–87.

Inhelder, B., & Piaget, J. (1958). *The growth of logical thinking from childhood to adolescence*. New York: Basic Books.

Lévi-Strauss, C. (1966). *The savage mind*. Chicago: University of Chicago Press.

Mbiti, J. S. (1969). *African religions and philosophy* (2nd ed.). London: Heineman Educational Books.

Odora Hoppers, C. A. (Ed.). (2002). *Indigenous knowledge and the integration of knowledge systems: Towards a philosophy of articulation*. Claremont: New Africa Books.

Ogunniyi, M. B. (2004). The challenge of preparing and equipping science teachers to integrate scientific and indigenous knowledge systems for their learners. *South African Journal of Higher Education, 18*(3), 289–304.

Posner, G. J., Strike, K. A., Hewson, P. W., & Gertzog, W. F. (1982). Accommodation of a scientific conception: Toward a theory of conceptual change. *Science Education, 66*, 211–227.

TIMSS. (2003). *Highlights from TIMSS: The Third International Mathematics and Science Study*. http://nces.ed.gov/pubs99/199081

Toulmin, S. (1972). *Human understanding* (The collective use and evolution of concepts, Vol. 1). Princeton: Princeton University Press.

Wober, M. (1969). Distinguishing centri-cultural from cross-cultural tests and results. *Perceptual Motor Skills, 28*, 488.

Part III
Medical Education and Practice

Chapter 5
Challenges of Medicine Across the Cultural Divide

Abstract Three incidents concerning caregivers and patients from different cultural groups illustrate medical challenges. A brief history of medicine in Africa during the colonial era illuminates some of the difficulties. I introduce the concept of medical pluralism and related ways of thinking as a synergistic solution. Finally I offer an approach to an integrative medical encounter.

The world population now stands at more than seven billion, and the developing countries comprise two-thirds of the world's population (Geohive 2013). On the whole, this means that western medicine is accessed by a minority of the world's population, leaving the vast majority to the practices of indigenous medicine. Yet the field of western medical has distanced itself from such practices, regarding indigenous knowledge (IK) as medically out-of-bounds. Currently there are some westerners who interface between the western medical system and indigenous knowledge systems (in the form of complementary and alternative medicine) without any cognitive complications. There are, however, many non-western people who are entirely happy to use both IK and western medical systems.

A large part of the world is under-developed but desirous of modernization and a better quality of life. There is a continuing demand for the developed nations to share their expertise, and alleviate some of the hardships and suffering of people in under-developed nations. This is particularly true for healthcare, where deaths are nearly five times more common than in developed nations, e.g., from childbirth, infections, and numerous treatable diseases.

Both donors and recipients consider modernization to be good and necessary. There are, however, serious costs to modernization. Forty years ago Roszak (1972) wrote:

> Science is not, in my view, merely another subject for discussion. It is **the** subject. It is the prime expression of the west's cultural uniqueness, the secret of our extraordinary dynamism, the keystone of technocratic politics, the curse and the gift we bring to history (Roszak 1972).

Part of this curse is that non-westerners have different expectations concerning health care services. These differences are reflected in patients' lack of satisfaction with care received from western medicine, for example concerning the lack of connection with the doctor, or lack of attention to the psycho-spiritual components of ill health. At the same time, western physicians' satisfaction with caring for patients from different cultural backgrounds is relatively low. This is

© Springer Science+Business Media Dordrecht 2015
M.G. Hewson, *Embracing Indigenous Knowledge in Science and Medical Teaching*,
Cultural Studies of Science Education 10, DOI 10.1007/978-94-017-9300-1_5

because of perceived non-compliance by their patients, who often do not follow medical directions. These problems are thought to negatively affect medical outcomes (Cooper et al. 2006). When there is misunderstanding by either the physician or the patient, it usually reflects a difference in the values and beliefs held by each party. Misunderstanding between physicians and their patients can cause anxiety, distress and even anger in the patient. There can also be disappointment and frustration for the physician who finds his care has been less than adequate, for example, when patients do not follow the physicians' advice or fill the recommended prescriptions.

With their specialized scientific worldview, Westerners (Americans and Europeans) are hegemonic—a powerful but small society dominates a much larger, culturally different, but less powerful society. This hegemony affects and changes the less-powerful people's values and beliefs, explanations, conceptions and perceptions, as well as hopes, fears, and emotions (e.g., expectations of happiness).

Currently, the USA is the most culturally diverse nation in the world. The term melting pot has been used to describe this phenomenon. But actually, there is no true melting together of the diverse groups but a cultural fruit salad, in which each group retains its particular cultural characteristics. There is an increasing need for cultural competence, i.e., being able to relate to diverse cultural groups in delivering health care in the USA. It is even more important in developing countries such as those in Africa, India, and Asia, where many people have indigenous viewpoints and subscribe to indigenous knowledge systems.

Cultural Misunderstandings in Medicine

This case report describes inter-cultural medical problems in terms of language differences and the medical ideas of an Egyptian Arabic patient and his family.

Case Report: Egyptian Visitor to the USA

Mr. Ahmed, a 28-year-old man of Egyptian birth, had come to the United States to join his mother, two brothers and one sister. He visited the emergency room of a university hospital complaining of a severely swollen and inflamed neck. He was in poor condition, and had difficulty speaking. He spoke English haltingly, but did communicate (with his brother's help) that three years earlier he had been told that he had cancer. Mr. Ahmed said that he had been treated in foreign hospitals with X-rays and with drugs. He had not seen a doctor since his arrival in the United States six months earlier.

The healthcare team noted swollen, rubbery lymph nodes in his neck and made a presumptive diagnosis of Hodgkin's disease (cancer of lymph nodes).

(continued)

(continued)

Mr. Ahmed was admitted for the treatment of his bacterial skin infection and the swelling of the neck, as well as for possible chemotherapy.

His family accompanied him and remained constantly in the visitors' room. The mother spoke no English. The brothers and the sister spoke it adequately but not fluently.

When Mr. Ahmed and his family were questioned about personal and family health history, and business or social status, they reacted with suspicion. As a result, many questions were not answered.

On clinical examination of Mr. Ahmed, the team recommended he should have lymph node and liver biopsies. Initially the family seemed to agree with the recommendations. But when the team asked Mr. Ahmed to sign the consent form, he refused. His mother and brothers strongly supported him and seemed very angry. Mr Ahmed refused to explain his reason for the refusal.

On the second hospital day Mr. Ahmed was sipping soup and choked. The team decided he needed a tracheotomy. As a result of difficulty in surgically placing the tube in his trachea, Mr. Ahmed had five to seven minutes without oxygen and entered a coma. The team started the patient on a respirator, but he had suffered serious brain damage and remained in deep coma. Although his neurological status did not meet the criteria of brain death, the neurology consultants confirmed a poor possibility for Mr. Ahmed's recovery.

Two team members (one spoke some Arabic) approached the family to explain Mr. Ahmed's serious situation as fully and carefully as possible. The interview was disrupted by extreme reactions of grief by the mother and by the family's attention to her rather than to the doctors. While the clinical team felt they had communicated the essential information, under these stressful circumstances, they did not yet feel they could bring up the issue of termination of life support.

A few days later the clinical team reached the conclusion that Mr. Ahmed was terminal and would never recover. It was their unanimous opinion that respiratory support should be terminated. The team announced this news to the patient's family. They were met with anger and outrage. The team consulted with the hospital's medical ethics committee and obtained legal counsel. Both groups supported the team's opinion that termination of life support was ethical and in accord with leading legal opinion.

The clinical team summoned a Moslem sheikh to comfort the family. The sight of the sheikh in the hospital room heightened the family's mistrust of the team. They became extremely fearful that, in spite of their pleas, their son/brother would now be left to die and no measures would be taken to save him. Suspicion increased and cooperation decreased. They felt the situation was a failure by the health care team. The family raised many angry questions and made the team members very uncomfortable. The damage from this well-intended act of compassion was profound.

Adapted from Meleis and Jonsen (1983)

This case reveals that the care team spoke English while the patient and his family spoke Arabic, with only a little English. This led to poor communication between the patient and his family and the healthcare team. More than language differences, this case exposes the negative consequences of the care team's lack of understanding of the cultural differences in the patient's (and his family's) values and attitudes. These misunderstandings negatively affected the patient's expectations and finally resulted in an unhappy clinical outcome. When these values and attitudes were disregarded, the patient and his family became alarmed, fearful and angry. In the medical world, such patients may be viewed as noncompliant, uncooperative, and even become labeled as 'difficult patients'—an unfortunate phenomenon that does not bode well for the patient.

While Arabs and other indigenous communities respect the powers of western medicine, they sometimes experience western medical practices as culturally discordant with their own views of appropriate caring behaviors (Meleis and Jonsen 1983).

Western medical practitioners are bound by ethical principles that provide guidelines for clinical practice, but may not mesh well with non-western patients, as in the case of the Egyptian patient. These include: patient autonomy (based on the idea that individuals are capable of rational decision-making, and should be able to take responsibility for their own decisions); truthful communications (believing that untruths are deceptive and immoral); medical futility (a realistic acceptance of the limits of medical effectiveness in certain circumstances); confidentiality between the doctor and patient; truth-telling at all times; avoidance of conflict of interest between physicians and economic entities such as pharmaceutical companies; and informed consent.

Western medical practice insists that patients must sign informed consent documents that give permission for medical procedures or treatments to take place. This is a legal requirement that serves to protect doctors from litigation as well as from medical mistakes. Arabs appear to find the requirement of written informed consent before surgery or surgical procedures as problematic. Meleis and Jonsen (1983) report that most Arabs believe that a verbal agreement is sufficient. Insistence on signatures on numerous hospital forms is seen as offensive because it suggests distrust of the patient's word of honor.

Arabic patients do not respond well to the direct disclosure of a serious diagnosis or a poor prognosis. While they respect western medicine, they do not believe negative information is helpful for a patient. Arabs, Africans, and other non-westerners view confronting a patient with his bad diagnosis and imminent death to be tactless and unforgivable.

Both Arabs and Africans believe that care giving should continue until the patient dies, according to the will of Allah (or God, or the ancestral spirits). They are especially resistant to decisions to terminate care. Instead, they believe that hope is essential to help the patient help him/herself. To do otherwise would be to give up on the patient, and is tantamount to issuing a death sentence. In southern Africa, traditional healers (THs) and their patients believe that when death is near it is the spirits who call the patient to the next life.

A group of respected family members often serve as the clearing house of medical information. In Africa, equivalent groups called therapy management groups consist of respected kin who serve as brokers on behalf of the patient (Janzen 1978). The task of the group of people who surround the patient in stressful times is to negotiate care decisions on behalf of the patient. Their job is to stay with the patient and fight for him/her, as well as to offer support and love. Caregivers need to speak through this group, which then decides what should be told to the patient, how, and when.

The role of the TH is to support the patient in the dying process, as well as the needs of the care management group and the family. They would be unlikely to talk of terminating care because that is the role of God alone. Neither Arabs nor Africans mention death except in highly euphemistic, metaphoric or ambiguous terms. Meleis and Jonsen (1983) suggest that Arabic patients expect humanistic care rather than what they perceive as cold, business-like transactions.

Both Arabs and Africans prefer to talk around a topic before coming to the point. Africans find it particularly rude when westerners focus directly on the purpose of the conversation before asking the *pro forma* questions about daily events, the weather, etc. These communication styles can be infuriating for westerners who value efficiency and succinct conversations. Meleis and Jonsen (1983) point out that these cultural differences are not only generally true for Arabs, but also for many other cultures.

Cultural Misunderstanding of Hmong Refugees in the USA

In Anne Fadiman's (1997) book: *The Spirit Catches You and You Fall Down,* she described the tragic case of a Hmong child from a refugee family in California who was afflicted with severe epilepsy. The well-meaning western caregivers and the adoring parents came to an impasse over the care of this child. The parents believed their child was imbued with special spiritual powers—the kind that qualifies a person to become a shaman (a person with healing powers). They sought the services of a local Hmong shaman to care for their very ill child and faithfully followed his advice concerning a sweat lodge, animal sacrifices, and other procedures.

For the Hmong, the meaning of epilepsy and the role of shamans in healing this (and other diseases) is similar to those of Africans who see epilepsy as a sign of special spiritual powers. People suffering from epilepsy and other psychological and neurological problems often become traditional healers as a response to the call of the ancestral spirits.

Meanwhile, in this Hmong case, the western care team planned to treat the child with conventional epileptic drugs. Their stress about this case was intense— they were doing their best. They believed the parents were noncompliant and felt that the child would not get appropriate care if she remained at home. At one point they managed to get a court order to remove the child from the family and put her in foster care for ten months. This resulted in great suffering for the child and especially her parents. Later on the child had several ongoing problems plus she suffered septic shock and fell into a coma. The care team decided that her recovery was impossible and recommended termination of all life support. They believed they had the parent's agreement for this, although this was not clear. After uncoupling the feeding tubes, the child did not die, but survived as a mere shadow of a person.

Although western ethical principles were probably not breached, it is likely that Hmong sensitivities to the role of a sick person's spirit and soul were given little attention. The need for this mother to comfort and care for her child was negated when the child was put in foster care. And when the care team terminated her life support, the family experienced a profound breach of trust. Yet the child lived on.

The cultural divide between the family and care team was huge. Fadiman summed up the situation: 'I have come to believe that (the child's) life was ruined not by septic shock or by noncompliant parents, but by cultural misunderstanding' (1997, p. 262).

Cultural Complexity in Lesotho

During my time as a lecturer in science education at the University of Botswana, Lesotho, and Swaziland, I was also in charge of a residence for women students. I lived in the residence, which allowed me close contact with my students. They came from all three of the southern African countries to study at the university. Many of them came from rural living situations.

Sick Student at the University, 1968
Someone was knocking at my door in the women's residence at the university. It was 2.00 AM. I was sure I was dreaming and turned over to sleep. But no, the knocking continued. I roused and stumbled to the door that opened to the long corridor of the residence. Two students were waiting.

"She's sick," said the one, looking at her friend. "She needs to see the doctor."

She told me that Ntombesan had a stomach pain and she seemed to be in distress. I knew that severe stomach pain is always a potential cause for concern. They were already dressed so I dressed too, and we went to find my old Volkswagen Beetle in the parking lot.

The African night was humming with the sounds of insects, frogs, and night birds. The sky was peppered with twinkling stars. The Milky Way and the Southern Cross were clearly visible. There was no moon that night, making the stars that much brighter.

(continued)

(continued)

The mission hospital was in another village not far from the university. Two missionary doctors from Belgium ran a small clinic and an even smaller inpatient hospital. Eventually a sleepy doctor with tousled hair opened the door.

The doctor took Ntombesan to an examining room while we waited on the hard metal chairs in the desolate passage. When she returned, the doctor drew me aside to talk privately.

"I can find nothing here, *pas de tout*. So I am giving her some pills. They are placebos. Bring her back tomorrow if she is not better?"

The candor of this doctor surprised me. Was she used to situations of unexplainable pain, and had found that placebos were effective? Could it be that this uncertain diagnostic situation was common for western doctors serving in the African environment? I was disappointed yet relieved that the doctor found nothing seriously wrong.

Early next morning I went to check on my sick student. She wasn't in her room. So where was she? Perhaps she was better already. I asked another student if she knew what had happened to Ntombesan.

"She's gone to town by the hitch-hike. She is looking for the traditional healer."

OK, OK, calm down now, I told myself. Let's see what happens now. My anxiety climbed—I was responsible for these girls. I went about my lectures and duties, nervous because I didn't know where she was or the state of her illness.

Later that day I found Ntombesan on campus, looking good, without stomach pain. She was reluctant to talk about what occurred on her visit to traditional healer.

This case perplexed me. I could not understand why the student, after I had taken her to the hospital, had decided to go and see a traditional healer (TH), at considerable expense to herself. At the time I did not understand the role of THs, and I felt the student should have told me what she planned to do. My understanding of THs was confused with the evil reputation of witchdoctors. I had little idea of the benefits of THs in the community. I worried that nothing good would come of this visit. I knew that rural African students found life at the university stressful, as there were so many academic expectations of them. In addition, life in the dormitories, and the food in the cafeteria were different from home life. In my mind I attributed the student's symptoms to stress. My ignorance of the African way of knowing contributed to my lack of understanding of the needs of this student.

Several years later I asked a TH to help me understand this event. She explained that the placebo probably did not work for this student otherwise she would not have looked for a different type of help. The student's problem could have been a type of colic, for which there are two possible causes: either a natural cause, or the intrusion of a bad spirit. Both types of colic cause stomach aches. The TH would have divined the cause of the problem, and then prescribed appropriate herbs. Presumably the TH diagnosed an intrusive bad spirit had caused the colic in the student, and cured her with appropriate herbs (Hewson and Javu 2011).

Colonial Medical Care in Africa

Africans in colonized countries found that the colonizers offered plenty of interest to the locals. Tools, beads, guns, metal cookware, blankets, clothes, etc. were all seen as highly desirable commodities. Africans were also intrigued with the curative capacity of these foreign people. To a great extent, the colonization process proceeded peacefully, with the local Africans eager for western goods and techniques. The colonists traded their offerings in return for furs, exotic wood, ivory, hunting opportunities, and eventually, control of the land itself.

African traditional medicine is rooted in social, cultural, ecological and historical circumstances of the people (Janzen 1978). This system of healing has existed for thousands of years. Although all knowledge systems evolve over time, the fundamental principles have remained the same. For example, African herbalism concerns the healing power of plants. This power is subordinate to the mystical powers of the spiritual world. The healing powers of plants can be used to affect both the physical bodies of individuals as well as the social harmony of a group (Janzen 1978).

In many ways, African healing systems seem similar to those of the Native Americans (North and South American), the tribal people of Asia (Malaysia, Thailand, Burma, Siberia, etc.), and the native people of Australia and New Zealand. This system rests on the premise of the power of God and spirits to provide healing, solace and good fortune.

Colonizing powers in Africa (for example, English, Belgian, Dutch, Portuguese, and German) became interested in providing medical services in Africa early in the twentieth century. At first, the focus of medical interest was on afflicted westerners who were trying to survive in the dangerous, unknown African environment. These dangers included animal bites (especially snakes), parasites (such as worms and *Bilharzia* or snail fever), infections (such as malaria, sleeping sickness, as well as *E. coli*), and wound sepsis. Later, medical attention also turned to relieving the illnesses of the Africans themselves, especially physical ailments, toothache, childbirth deaths, indigenous diseases such as malaria and sleeping sickness, and a vast number of other illnesses.

Africa became known as 'the sick continent.' Yet colonists themselves unwittingly introduced a large number of these diseases, such as smallpox and rinderpest (a type of plague affecting cattle), as well as infectious diseases such as measles, mumps, rubella, whooping cough. This label of 'the sick continent' seems, in retrospect, unfair (Prins 1989).

Early biomedical theorizing concerning Africans gave rise to a prevalent colonial view: the colonization of Africans to western ways of thinking led to separation from their own indigenous ways of knowing—and this led to their disintegration and mental illness. According to this theory, western civilization was making Africans crazy (Vaughan 1991, p. 108). This particular theory has had deep and abiding consequences for white colonial perceptions of black Africans—it fathered a racist view of the capacities (or lack of them) of Africans when compared with westerners. This view survived many generations up to the recent past of South Africa, with the establishment of *Apartheid* and the second-class system of education for African children.

Sacred Versus Secular Colonial Medicine

Medical missionaries initially took on the role of introducing western medical practices in Africa. They insisted that their medicine combined both scientific medicine and belief in the supernatural (Vaughan 1991). Because of this, mission medicine had some important similarities with traditional African medicine—both insisted on the supernatural power of a deity and spiritual influence of beings (the Holy Spirit in Christianity and ancestral spirits in traditional African cosmology). Medical missionaries referred to God as the 'great healer in the sky, and the healer of souls.' Mission doctors and nurses also insisted on Africans' adherence to Christian beliefs and practices. It was a prerequisite for providing medicines and procedures to African patients, and also served the purpose of (apparently) promoting the spread of Christian doctrine in 'darkest Africa.' It was not too difficult for Africans to go along with medical missionaries as long as these foreigners did not interfere with important components of the African traditional way of life, such as initiation ceremonies.

This was not the case for secular medicine. Africans found little in common with the principles of western medical science, except that medicines (especially injections and pills), tools (such as X-rays and tooth extraction devices), and procedures (such as surgeries with anesthesia and childbirth methods) seemed persuasively effective. The Africans saw that the white men's medical offerings were powerful, especially in terms of these concrete ailments and had no difficulty taking advantage of the healing techniques that were offered to them. But at the same time they adhered to their African understanding of illness involving God and spirits (good and bad) and their approaches to maintaining and regaining equilibrium and health.

The Western Medical Way of Knowing

Western medicine is a branch of science and increasingly conforms to scientific principles. The prominence of evidence-based medicine is testimony to this evolution. Western medicine holds that empirical data gained from randomized, controlled clinical trials is the benchmark of medical decision-making and treatment. Medical principles rest on several scientific axioms—starting points for reasoning. One of these is Occam's razor, concerning the principle of parsimony, and succinctness in problem-solving. It states that among competing hypotheses, the one that makes the fewest assumptions should be selected. Thus one theory should explain and predict as many phenomena as possible, and as simply as possible. Medical scientists also pay attention to the principles of generalizability and internal consistency when making decisions about illnesses and treatments. Moreover the criteria of validity and reliability must be met adequately in medical experiments. It would be inappropriate to have disparate theories for numerous phenomena. It would be even more inappropriate and irrational to have more than one theory for the same phenomenon. But these kinds of reasoning are found in African traditional medicine (see below).

Western medical approaches are reductionist—the focus is on individual parts or sections of the body (e.g., eye, digestive system, tissue cells, and DNA). These foci have created a highly developed, sophisticated knowledge system and effective clinical treatments. This reductionist approach results in the human body being viewed in terms of significant parts, with medical specialists focused on each. For example: the heart (cardiologists), lungs (pulmonologists), glands (endocrinologists), and so on. This reductionist approach means that western practitioners seldom see the human body as a whole. Moreover, the human body is seldom seen within the context of society and the cultural group, except in the relatively new fields of environmental medicine and public health.

Western medical practice views the human body as separate from the mind and soul. The focus on the material body excludes the spiritual components. In a confrontation between early scientists and the 17th century church (an academic turf war), Descartes convinced the church leaders that scientists would limit their research to the material world, leaving spiritual matters to the church. Materialism was thus a response to this Cartesian split between the body and mind. Western medicine is concerned with pathology and pathophysiology, which reflect the central idea that illness is caused by a malfunction of body systems or infection by extraneous organisms. While contemporary medicine is taking increasing interest in mind-body medicine, there is still no role for spirit.

The western medical approach seeks solutions to parts of the problem, e.g., through surgeons, physicians of all types, social workers, and psychiatrists. Medical treatments are based on the rigorous science of clinical trials in combination with advanced statistical treatments of the data. Experimental criteria such as reliability, validity, and generalizability prevail as the signposts of safe medical practice. Medical ethics has become important in maintaining good clinical practice, and patient-centered care giving is becoming increasingly important. Evidence-based medical practice is the contemporary standard of care, leaving no opportunities for physician intuition or hunches.

Donald Schön (1987) noted that in the professions (architecture, engineering, medicine) there is a difference between the high ground of theory and orthodoxy, and the messy world of practice. There is a zone in which problems may include uncertainty and conflicting values, and where modern medical standards of evidence-based practice are not possible. This is the 'indeterminate zone of practice' that seems similar to the African world, where ways of knowing are complex, idiosyncratic and different from orthodox western medicine.

Ways of Knowing in African Traditional Medicine

A traditional healer in Zaire (Nzoamambu) illustrates an indigenous medical viewpoint, which is similar to the viewpoints of southern Africans:

> The heart is the focus of a person, for the heart embodies the principle of life. Other organs and functions, such as the abdomen, head, blood, and thoughts carry on an existence partially independent of the heart. Affliction introduces anger into the organs and functions, causing their impairment and congestion. If localized treatment is ineffective in restoring health, an affliction worsens along a predictable hierarchic progression of symptoms and diseases for each organ and function until the heart—the lord of the person—is also adversely affected. However, when the heart is touched by affliction, diagnosis must look beyond the body for an explanation, to spirit- or illness-invoked anger, a sufferer's own carelessness toward moral codes, or witchcraft emanating from the social environment. These secondary and tertiary dimensions of heart, society and cosmos must be diagnosed and treated before an individual's health can be restored (Janzen 1978, p. 156).

To explain sickness and bad luck, this way of seeing illness integrates the person, the society in which he lives, the natural world, and supernatural beings. In general, traditional African medicine does not distinguish clearly between cause, diagnosis, treatment and cure, as in western medicine (White 1995). For example, African patients might complain that: "The sickness is in my feet" (meaning he has stepped on an evil spell); "The spirit of my ancestor kissed me" (meaning that super-natural forces are making him ill); "I have snakes in my belly" (meaning there is a spiritual intrusion that is affecting him).

African traditional medicine also does not subscribe to the principles of internal consistency or generalizability. Testing of diagnoses (for validity), tests (for reliability), and treatment modalities (for efficacy) are not part of African care giving. Traditional healers receive medical knowledge concerning a particular patient (diagnosis, selection of tests, treatment plan) from God through the ancestral spirits. This information should be believed not doubted. The spirits must be obeyed, not tested. For westerners, African traditional medicine and accompanying IK reveal a confusing and sometimes impenetrable logic.

Preparing for Death at the Mission Station, 1976

A remote Presbyterian mission station in the mountainous region in South Africa now known as Basotho Qwa Qwa has existed for many years. Over several generations the small station became a center of protestant Christianity, with devout practices and dedicated leadership. The mission church supported community outreach, with an orphanage and home for severely handicapped children. Many of the local Basotho people experienced meaningful Christianity there, and took on leadership roles within the church.

(continued)

(continued)

I visited the mission station and over a cup of tea the white minister told me the story of the Basotho man who became a deacon of the little church—a lay minister. He eventually became old and unwell. The minister suggested he should 'put his life in order' and make preparations for his foreseeable death. This did not seem to disturb the old man too much. The minister was impressed, and thought the old man was displaying the peace of mind and courage befitting a man of the church.

But then the old man announced he was going to visit the local traditional healer. The minister was horrified—it seemed a thoroughly un-Christian idea. He tried to dissuade the old man who held fast to his plan. The old man explained that he was definitely a Christian and believed in Jesus and the life hereafter—but there were issues that made him anxious. What would his ancestral spirits think of him if he did not pay them his respect? It seemed to the old man perfectly natural that he should visit the traditional healer and perform the prescribed rituals in order to make things right with his ancestral spirits. He also believed in his spiritual destiny through his faithful adherence to Christian practice. For him, both approaches seemed appropriate and sensible.

But the minister did not see it this way at all. Either a person is a Christian or he is not. You can't have it both ways. As he told me this story he shook his head sadly, noting that the old man's Christianity must be only skin deep. His return to the un-Christian practices of Africa seemed like a betrayal.

Visit to Basotho Qwa Qwa, South Africa (1976)

Negotiating Between Different Ways of Knowing

Over the past three decades, researchers have become increasingly interested in indigenous ways of knowing. This interest is particularly strong in education, where attempts are being made to educate different cultural groups in heterogeneous classes. The different ways of knowing are clearly manifest in science classes as well as in medical teaching. In order to reconcile different ways of knowing, cultural groups that have experienced prejudice and persecution by dominant groups who invaded and subjugated them need healing. There also needs to be rebuilding of indigenous people, some of whom are in danger of losing their indigenous ways of knowing. This will require bridging between indigenous and western knowledge (Battiste 2000, p. xvii).

To create a bridge between the ways of knowing of different cultural groups we need to examine the ways of knowing of indigenous people and identify the differences and similarities between them. Aikenhead and Ogawa (2007) describe some of the most salient differences between IK and western science and report on numerous efforts to enhance the cultural survival of indigenous people worldwide (see Chap. 2).

Successful interaction between two cultures with different epistemologies requires dialogue and mutual respect between the representatives of the different ways of knowing. The outcome of such dialogue and mutual involvement can be seen as a type of synergism in which the net consequence is a type of knowledge system of which each group may be independently incapable of creating (Hewson 1988). Examples of synergism may be seen in terms of different ways of thinking such as medical pluralism, kluge, *bricolage*, syncretism, entanglement, cherry-picking, and equipollence. The table in Chap. 2 distinguishes between western science and indigenous knowledge. Based on these fundamental differences between the two ways of knowing, it is possible to see that various hybrid approaches to understanding the world and ways of thinking could occur. These approaches are not equivalent, nor are they discretely different. They do provide a spotlight on different aspects of synergistic thinking.

Medical Pluralism and Equipollence

Medical anthropologists have observed types of thinking where Africans have accepted and adopted aspects of western medicine that suit their needs, without accepting the whole medical game plan with its axioms and intellectual rules. This type of parallel thinking is referred to as medical pluralism (Vaughan 1991; Whyte 1989; Janzen 1978).

Despite western disparagement of this type of thinking, it is common in non-western societies, and it also occurs within western societies. For example, a western medical colleague maintains a strict medical model in her practice. But when she herself needs medical treatment she is open to both western medicine and alternative (complementary) medical practices. She refers to this situation as being in an 'epistemological bind.' This type of bind seems to be a western problem, but does not appear to be problematic for non-westerners.

Several types of deviations from orthodox scientific thinking have been described in recent years. They suggest that 'shades of grey' exist between formal scientific thinking and everyday thinking. I propose that these could be described in terms of epistemological pluralism—of which medical pluralism is an example. To illustrate these ways of thinking:

Kluge—a clumsy or inelegant yet surprisingly effective solution to a problem (Marcus 2008, p. 2). Kluge involves a certain amount of higgledy-piggledy or haphazard thinking where the end product justifies the means of getting it. The end product to this kind of problem solving may be 'good enough,' yet not necessarily perfect or elegant. Kluge can commonly be found in engineering tasks, or in the way plumbers or electricians accomplish their tasks of fitting pipes or wiring electrical circuits in a house. Evolution involves a process of kluge where the end-state is survival of the species. The anatomy and physiology of the human body reflects evolution and many of the workings (such as the structure and function of the eye) involve kluge—strange and illogical structures. Even the human brain is the product of haphazard evolution. Marcus (2008) describes several kluge bugs in our cognitive makeup—we are not perfect creations.

All ways of knowing are somewhat imperfect and subject to kluge, and indigenous ways of knowing are probably more susceptible because they are subjective, holistic, monist, relational, idiosyncratic, deterministic, spiritual, anthropomorphic, and so on. On the other hand, the western way of knowing is causal, reductionist, objective, mechanistic, probabilistic (etc.), and has less kluge. As an example, computer sciences of information processing have little kluge because they function strictly on the basis of exact, nearly perfect logical systems that adhere closely to the western way of knowing.

Bricolage—the construction of ideas using whatever is available, producing a 'mix of this and that,' also referred to as 'soft' skills for understanding the world (Marks 1997, p. 214; Lévi-Strauss 1962, p. 16). For example, in cases of clinical uncertainty, a clinician may use the approach of testing for a wide range of possible pathologies, patho-physiologies, or infections, believing that something will show up in the panel of tests (Hewson et al. 1996). This approach may not be efficient, but it works, and is currently used in medical practice in the USA.

Syncretism—the combination of different forms of belief or practice such as the fusion of two or more originally different forms. The above narrative of the Christianized Basotho man may be seen as a syncretism between the traditional African worldview and Christianity, where both views are seen as important. For many non-westerners, it is not a question of 'either this or that' but rather 'both this and that.'

Entanglement—when a non-western community (such as African) takes some aspects of western medicine, unpacks and examines them, and then reinterprets them in such a way that they are adapted to their own situation, giving the conceptions local meaning (White 1995, p. 1395). For example, White describes how in Uganda, African patients particularly appreciated stethoscopes, X-rays, and injections. In addition, they fancied pink pills. These accoutrements of medicine were adapted as locally acceptable healing agents with great power. The pink pills, in particular, were seen as powerful because the pink color was viewed as 'hot' and useful for dealing with illnesses caused by 'cold.' In this way, medicines were dispensed more as charms rather than as medications with active healing properties. Ugandans, like other Africans (see Chap. 4) have an etiology of disease in which concepts of 'hot' and 'cold' are widely applied to illness and health. Mixing and matching western and non-western medical conceptions is common.

Cherry-picking—selective use of facts or points in an argument in order to create a point of view, while ignoring other facts that do not support the point being made. This may lead to magical thinking, where certain ideas are emphasized and generalized to the exclusion of others. For example: in current western medical practice, evidence-based medicine is highly encouraged in clinical diagnosis to the exclusion of doctors' anecdotal evidence or their clinical hunches. Marks (1997) described how Africans accepted the components of western medicine that appealed to them, and overlooked those that did not suit them, both conceptually and practically. For example African interest in pills was mainly focused on red or pink pills, which were seen as powerful, even magical remedies. According to Marks, 'colonized people use an indigenous medical practice that reflects selected incorporation of, adaptation to, and manipulation of western medical practices' (Marks 1997, p. 214). Cherry-picking may also involve clear resistance to certain western medical ideas. For example, Africans rejected colonial attempts to stop initiation rites for boys (because of the frequent infections after circumcision). They also rejected the idea of western doctors and nurses to not allow therapy management teams consisting of elderly women to attend the birthing of babies (because they were seen as dirty).

Equipollence—happens when two competing ideas exist at the same time in a person's mind. The two ideas may differ significantly, but in the mind of the individual they have equal power, effectiveness, and significance. Ogunniyi calls it harmonious dualism or equipollence (2000). The above story of the old man preparing for his death reveals a common phenomenon in Africa—the ability to hold two or more ideas at the same time, even if they conflict. Westerners, such as the mission minister in the above narrative, would be likely to interpret this phenomenon as peculiar and irrational.

Meshach Ogunniyi in the School of Science and Mathematics Education, University of the Western Cape, South Africa became deeply interested in educational philosophy and practice. Originally from Nigeria, he trained as a physicist, and then followed a productive career in science education. I interviewed him about the co-existence of competing ideas within a single individual.

Equipollence

"The current educational emphasis concerning indigenous knowledge, at least in Africa, is to teach science that relates to the student's socio-cultural interests. We need to see how science and indigenous knowledge complement each other. We know that students have a lot of misunderstandings concerning school science. But I propose that we need to accept that indigenous and scientific knowledge can co-exist within an individual in a meaningful way. Trying to change students' traditional ideas of the world to the scientific view is just like pursuing the wind.

I have been saying since the early 1980s that there is no way you can take a person out of his own experiences otherwise he is going to lose his sense of cultural identity. You can teach a person about modern science as much as possible, but you have to also allow him to rely on the existing ideas of his life that make sense to him, because it appeals to his interests and his fundamental understanding of the nature of the world, his sentiments, and his culture. These different things essentially make up who he is as a person.

I think that if there is any conflict between ideas (indigenous knowledge and western science) it is temporary because the individual always finds a way of adjusting the conflict, through what I call integrative reconciliation, accommodation, and finally adaptation to the new worldview. Otherwise there is no way for the student to learn. We can say to the student: 'OK, you bring all this experience and knowledge from home. This is the way science looks at it. So how do you make sense of it?'

(continued)

(continued)

In my own study of science (as an African) I have never experienced the conflict that I've seen described in papers. Learning science does not make me lose my indigenous knowledge. I just see that indigenous knowledge and science are playing different roles. When I go home I do all the things that traditional religious Yorubas do. When I am at the university I think scientifically. We all play different roles on a daily basis—as a scientist, a teacher, a father or mother, or as a husband. I do not construe the differences between the two knowledge systems as conflicting. Rather I see them as having different but complementary roles.

Of course I have found Niels Bohr's complementarity principle in physics concerning wave and particle theories very useful. So what I propose is harmonious dualism between two different conceptual explanations. Harmonious dualism is not static, but it's a dynamic relationship, for example between my indigenous world where I live daily and the science I do at university. These views are related to each other—one is not greater than the other. Each one has its own essential useful role to play. In my recent papers I introduced the idea of contiguity theory as a kind of platform for explaining harmonious dualism. Then I introduced the term equipollence to describe the harmonious co-existence of two different views, theories, or conceptions.

I don't see science and indigenous knowledge as dichotomous as most people do. Science addresses empirical questions, but it's highly inefficient in addressing other things such as explaining the fundamental nature of being. There is no point in dichotomizing a body of knowledge that includes our tripartite nature— body, soul, and spirit. The body is where we can see and touch, the soul is where the mind is, where we make decisions, and the spirit is something that is beyond the human level. We deal with spirit through prayers and meditation. These are all components of our human existence."

Interview with Dr. Meshach Ogunniyi (2010)

This interview illustrates how Ogunniyi accepts the epistemologies of science, and African traditional thought and spirituality as equipollent—he has never experienced conflict between them. For an African, this is not surprising. However, westerners often do find the divide between these epistemologies troublesome.

In the previous narrative about the mission station, I showed how African interest in and acceptance of Christianity did not necessarily involve a rejection of one epistemological commitment in order to live according to another. It was an example of equipollence.

Integrating Western and Non-Western Approaches

Synergy is obtained when two or more agents work together to produce a result not obtainable by any of the agents independently. Can the western medical approach work alongside African traditional healing approach? Could there be a synergistic

Fig. 5.1 Mirranda Javu,
Traditional Healer, Western
Cape, South Africa
(Photographer: Peter
Hewson)

relationship between the two approaches? From the western view, it may seem that the two approaches are incompatible and therefore incommensurable—lacking any common factors that would make the two approaches able to work together.

One of my informants is a traditional healer named Mirranda Javu, who lives in Phillipi East, near Cape Town, South Africa. She is from the Xhosa tribe and speaks both Xhosa and English. She is a student at the University of the Western Cape and works for the South African Medical Research Council (MRC) as a research assistant in the Indigenous Knowledge Systems [Health] Lead Programme. This program is for identifying indigenous knowledge (mainly herbal) concerning health. She interfaces between traditional healers and western researchers (Fig. 5.1).

I asked Mirranda Javu about her views on working with both IK and western medical approaches. I wanted to know how she served as a traditional healer at the same time as being as a research assistant at the MRC.

Mirranda Javu on Integrating African and Western Approaches

"I manage both approaches very well. In my job I gain a lot of experience and a lot of knowledge about traditional plants. This has contributed to my healing skills. I learn so much, for example, the botanic and English names of plants, and the medical names of illnesses and diseases. My job is what is in me. Working with communities is what I like most, especially doing something that is in line with my traditional healing. I facilitate and conduct TH training workshops on western views of illness and disease, I educate school learners about traditional plants, and I promote awareness on how the Medical Research Council handles TH's claims for indigenous cures and treatments. It is not disturbing at all. But it's exhausting—that I must say.

I use both approaches myself. I use western medicine because I do visit western doctors though I can also use plants in my shrine. If I have a headache I take a *panado* (over-the-counter analgesic) because it makes me feel better. But I also know which plants I can use for a headache. I can take a specific plant and burn it

(continued)

(continued)

and then snuff it, and my headache will disappear. I take both approaches. I grew up visiting western doctors. There is nothing wrong with that, and I cannot just wipe that out of my mind overnight. I believe in traditional medicine—I strongly believe in it. I know and I am sure that traditional medicine works wonders. I know that because I use traditional medicines when treating my clients and I have proved it works. It is something that I cannot run away from—it is true.

The advantage is that I understand both systems. Also, I know when a headache is not suited for *panado*, in which case I must do something else. I can find out what is the cause of a person's headache. We THs believe in finding out the cause of a problem. So I can give you a plant to search the cause of a headache through your system without divining, but I cannot give you medication for headache when the problem is in the stomach. That is why it is important for us traditional healers to know more about the human body, the anatomy of the body. For example, if you have got a headache, what is it that makes you vomit? The western doctor would say 'You have a migraine,' or something else using terms that I may not understand. But I would say that you have got a poison in your stomach, maybe due to evil spirits. And you find out this poison in the stomach might be the cause of the headache. So in that case you are not supposed to treat the headache, you must treat the poison.

I check for the cause of the problem through divination. That is what we diviners call 'a gift.' If you don't have that gift you won't know the cause. But sometimes if I suspect the cause from the signs and symptoms I just prepare a special medication and give it to a client to drink, and the medication will check for me. The medical doctor would not have the same understanding of some causes that are not natural, and possibly due to evil spirits. So that is why you have to check the cause by searching or divining. So the advantage for me is in understanding both systems. It only needs knowledge of how to differentiate things.

So these two systems do work together for me. I do not experience any negatives. Times are changing you know, and this Medical Research Council experience helps. My medical knowledge helps me in terms of diagnosis and referrals. And it will help me in the future as well. At least I will understand some of the terminologies and the different jargons that are used in both systems."

Interview with Mirranda Javu, Western Cape (2010).

The inspection of differences between African traditional medicine and western medicine is revealing. Westerners experience the conflicts between the two epistemologies as disturbingly problematic—they see internal conflicts and irrational thought processes. African traditional medicine seems to be bad medicine. It is no wonder that Colonial powers tried to eliminate traditional medical practices in numerous countries: Australia, North America and Africa. But Africans see a different picture—if a medical practice works they will accommodate it. They will also adapt it so that it fits with their worldview. This could be labeled as any one of the above descriptions of knowledge integrations: medical pluralism, kluge, *bricolage*, syncretism, entanglement, cherry-picking, and/or equipollence. From a western viewpoint it is unclear how the people in these three examples managed cognitive reconciliation of two different ways of knowing—IK and western medical knowledge. Apparently they did so, and were content and comfortable with their understandings.

Integrative Approach to Patient Care

I have proposed a four-step approach to an intercultural physician-patient medical encounter. These steps are: (1) prepare, (2) ask, (3) consider your options, and (4) tell the diagnosis and treatment plan (Hewson and Budev 2000).

Prepare for the Encounter. Greet the patient and introduce yourself, using formal titles. Then check your patient's familiarity with English to see if you need to use an interpreter. Ensure your patient is comfortable with you as a provider, in terms of your age and gender. Adjust your verbal and nonverbal language by using appropriate vocabulary and pace. Do not raise your voice or speak as if to a child. If using an interpreter, still talk directly to your patient.

Ask Questions During Clinical Examination. Use open-ended questions to establish the patient's background (country of origin, home language, etc.). Identify your patient's 'explanatory model of illness' in terms of his or her family system, and social, financial, spiritual and cultural contexts. Identify the patient's expectations of you as a care provider. Draw out your patient's concerns and fears, and check for his/her adherence to traditional or alternative medical practices. Help your patient describe symptoms using diagrams, drawings, dolls, etc.

Consider Your Options. Decide how you will respond to your patient's cognitive and affective states, as well as his/her medical needs. Decide on the best way to handle this encounter by negotiating the agenda for the visit; resolving possible misunderstandings; and encouraging therapeutic compliance.

Tell the Diagnosis and Treatment Plan. Tell your patient the diagnosis in the context of his/her expectations and background. Explain your reasons. Check the patient's understanding of the diagnosis and check his/her feelings about it. Respond to the feelings explicitly. Specify your treatment advice (what, how, when) and negotiate a treatment plan. Find out if the patient is willing to follow your advice.

We Can Teach the Children: Additional material to this book can be downloaded from http://extras.springer.com

References

Aikenhead, G. S., & Ogawa, M. (2007). Indigenous knowledge and science revisited. *Cultural Studies of Science Education, 2*, 539–620.

Battiste, M. (Ed.). (2000). *Reclaiming indigenous voice and vision.* Vancouver: University of British Columbia Press.

Cooper, L. A., Beach, M. C., Johnson, R. L., & Inui, T. S. (2006). Delving below the surface: Understanding how race and ethnicity influence relationships in health care. *Journal of General Internal Medicine, 21*, S21–S27.

Fadiman, A. (1997). *The spirit catches you and you fall down.* New York: The Noonday Press.

Geohive. (2013). *Global population data.* www.geohive.com/earth/population_now.aspx

Hewson, M. G. A'B. (1988). The ecological context of knowledge: Implications for learning science in developing countries. *Journal of Curriculum Studies, 20*(4), 317–326.

Hewson, M. G., & Budev, M. (2000). Cultural consults for caring trans-culturally. In *Medical encounter.* Publication of the American Association for Communication in Healthcare.

Hewson, M. G., & Javu, M. T. (2011). Indigenous knowledge concerning health, illness and medical education in Southern Africa. *Medical Encounter, 25*(2), 13–18.

Hewson, M. G., Kindy, P., Van Kirk, J., Gennis, V., & Day, R. (1996). Strategies for managing uncertainty and complexity. *Journal of General Internal Medicine, 11*(8), 481–485.

Janzen, J. (1978). *The quest for therapy: Medical pluralism in Lower Zaire.* Berkeley: University of California Press.

Lévi-Strauss, C. (1962). *The savage mind.* Letchworth: The Garden City Press.

Marcus, G. (2008). *Kluge: The haphazard evolution of the human mind.* New York: Houghton Mifflin Harcourt Publishing Co.

Marks, S. (1997). What is colonial about colonial medicine: And what has happened to imperialism and health? *Social History of Medicine, 10*(2), 205–219.

Meleis, A. I., & Jonsen, A. R. (1983). Ethical crises and cultural differences. *Western Journal of Medicine, 138*(6), 889–893.

Ogunniyi, M. B. (2000). Teachers' and pupils' scientific and indigenous knowledge of natural phenomena. *Journal of the Southern African Association for Research in Mathematics, Science and Technology Education, 4*(1), 70–77.

Prins, G. (1989). But what was the disease? The present state of health and healing in African studies. *Past and Present, 124*, 159–179.

Roszak, T. (1972). *Where the wasteland ends: Politics and transcendence in post-industrial society.* New York: Doubleday.

Schön, D. A. (1987). *Educating the reflective practitioner: Toward a new design for teaching and learning in the professions.* San Francisco: Jossey-Bass.

Vaughan, M. (1991). *Curing their ills: Colonial power and African illness.* Stanford: Stanford University Press.

White, L. (1995). They could make their victims dull: Genders and genres, fantasies and cures in colonial southern Uganda. *American Historical Review, 37*(2), 261–281.

Whyte, S. (1989). Anthropological approaches to African misfortune from religion to medicine. In A. Jacobsen-Widding & D. Westerlund (Eds.), *Culture, experience and pluralism: Essays of African ideas of illness and healing* (pp. 289–301). Philadelphia, PA: Coronet Books.

Chapter 6
African Healing and Traditional Healers

Abstract The healing practice of traditional healers is described in terms of: concepts of illness and health, the consultative process, diagnosis, treatment(s), and prevention. A traditional healer describes her healing practice. Incidents involving divination and the extraction of evil spirits are described. I offer a comparison between western and African healing approaches, and end with a note about African witchcraft.

African indigenous knowledge (IK) is concerned with health and healing, and is a well-developed indigenous science that has existed over thousands of years. The majority of African populations consult traditional healers (THs) for help with their problems—physical as well as psycho-spiritual.

Traditional African medicine concerns IK about health and healing. It explains and predicts phenomena in the domain of health and illness (such as: What is wrong with me? Why did I get sick now?), surviving and not surviving, happiness and unhappiness, power and a lack of power, success and failure, as well as good fortune and bad.

To fully understand the IK concerning African medicine, we must ask 'what' indigenous people know and 'how' they know it. In other words, we must focus on the ways of knowing of different indigenous knowledge systems—the paradigms, or prevalent shared frameworks of ideas for particular people in a particular context (Toulmin 1972). Paradigms are like conceptual goggles that largely determine the kind of things people think about, and the way in which they think about them.

People in different historical, geographic, linguistic, religious, and cultural contexts often have different paradigms. Individuals within a particular paradigm have a strong commitment to their own cultural ideas. Sometimes the differences between the western paradigm and those of other cultural groups are substantial, limiting communication about, or sharing of ideas. Other times the differences are small but irritating and perplexing—people may think they know what is being said or what is happening, but they don't. One of these perplexities can be found for western medical practitioners who depend on evidence-based medicine and standards of care, but do not find these same concerns in African care giving. Consequently many westerners view the African system of healthcare as idiosyncratic and the medical approaches as unpredictable.

© Springer Science+Business Media Dordrecht 2015
M.G. Hewson, *Embracing Indigenous Knowledge in Science and Medical Teaching*,
Cultural Studies of Science Education 10, DOI 10.1007/978-94-017-9300-1_6

This was the situation in the three examples described in Chap. 5. The cases of the Egyptian patient with Hodgkins disease, the Hmong child with epilepsy, and the African student with a stomach ache each involved working across the paradigms of non-western traditional healthcare and western medical science.

African traditional healers work in a paradigm, about which we know very little. We do know, however, that spiritual phenomena, e.g., the powerful influence of spirits from the recently departed ancestors, the existence of good and evil spirits, and the presence of spirit in both living and non-living entities are very real to traditional Africans as well as other non-western people (Hewson 1998).

We don't know much about Africans' ways of discovering or creating new knowledge, and their approach to discerning the truth or untruth of new information. This provides a challenge for medical educators wishing to improve the learning and teaching of western medical science to non-western students and patients. If we don't know the ways of knowing of our students and patients, how can we communicate appropriately with them? Conversely, if patients are not on the same page as their teachers and doctors, how can there be effective learning and healing?

Traditional Healers and Their Healing Practice

In many respects African THs are similar to European and Native American shamans who have existed for thousands of years, and represent the most widespread and ancient systems of human healing (Harner 1990). The shaman is known as the 'great specialist of the human soul' (Eliade 1964). While most contemporary studies have focused on shamans of North and South America and northern countries such as Siberia, there has been rather less documentation of African traditional healers.

Traditional healers (THs) exist throughout the African sub-continent and are represented in all the language groups. They have IK concerning health and illness, and ways of maintaining health as well as healing. THs have many roles such as doctors, counselors, psychiatrists and priests. They also manage birth and death in their communities. There are specializations within the category of TH, such as diviners, herbalists, birth attendants, and healers that 'cool' people who are suffering from spiritual heat.

THs believe in the bio-energetic body, as distinct from the biological body. This is similar to Chinese beliefs concerning the bio-energy of *chi* (involved in acupuncture), or Indian concepts of *prana*. In the language of the indigenous Khoi-San people of southern Africa, this bio-energy is referred to as *num* (Katz 1982). *Num* is everywhere. It originates with the divine being (God), and is found in relative amounts in ancestral spirits, humans, animals, plants, as well as the material of the earth (rocks, mountains, water, lakes, etc.). Healing plants are said to contain *num* (Katz 1982) and need to be treated with care and respect. This spiritual energy is the most powerful force in the experience of indigenous people in southern Africa.

African IK involves a central idea of ancestral spirits, the recently departed family members (parents, grand-parents, aunts, uncles). These spirits can, and do, influence the lives of the living (Mbiti 1985). These entities have the role of the 'living-dead'

(Mbiti 1985), and continue to exert influence (both positive and negative) in the lives of their descendants. Ancestral spirits are seen as intermediaries between humans and God. They are a source of power and knowledge. Africans connect with their ancestral spirits regularly, and ensure that they are placated through offerings and prayers.

THs help others in the community through providing insights about their problems. They offer healing through their extensive knowledge of herbal and animal-based medicines and therapeutic actions such as rituals. They also have knowledge of the traditional ways in which the community organizes and manages itself. Traditional healers in Africa have a powerful role in society through their easy access to the ancestral spirits (Hewson 1998). They often serve as advisers for leaders such as chiefs because of their knowledge of traditional matters and their perspicacity (Hewson 2011).

I interviewed Mirranda Javu, a TH from the Xhosa tribe who is also a graduate student in medical anthropology, and works for the Medical Research Council in South Africa as a research assistant in the indigenous knowledge programme. She told me about traditional healers and her own work as a TH (Hewson and Javu 2011).

Mirranda's Practice as a Traditional Healer

"Sometimes a healer may only be a diviner who cannot use plants for treatment. The diviner then has to refer the clients after divining to a herbalist.

Diviners need to find out if a problem might be *imvumisa*. This is where something happened and even though you were not there you have to know what happened. For example, if something really unusual happens (e.g., a dog comes into someone's house and urinates), a person will wonder why and ask what it means, and will go and seek help. As a traditional healer you have to be able to foresee the event. This is very difficult because you have to point out the event (without being told) and explain what the event signifies. Another example: an accident happened, and the person may have either passed away, or be in hospital. And the family will ask you 'Where is that person now?' And you have to be able to pick up that either he is in the hospital and still alive, or that he passed away. So *imvumisa* is very difficult—you need to pray to be assisted by your ancestors. You cannot just guess.

Then there is another type of divination called *ingxilongo*. This is for identifying human problems like headache, stomach ache, your sore leg and so on. If you are a diviner you can say these are minor problems. But they are not minor problems as such. If I tell you that you are suffering from stomach ache, but you actually have a headache, then I am a bad diviner. So *ingxilongo* should not be taken for granted.

There is also *unontongwana*. People will come when they are missing things—maybe the car has been stolen, or money is missing, or someone is

(continued)

(continued)

missing. So you must be able to tell them 'OK, you are here because your car is missing. Your car has been stolen by so-and-so.' They will ask you 'Where is it now? Can you get it back?' You should be able to know those things. But for me, no, I don't like that type of divination because people fight amongst themselves—it creates conflicts.

I am a diviner. I also do treat my clients, but I am not a herbalist. My skill as a diviner became obvious in my training process. When people come I usually know their problems, but sometimes I don't. When they come, I greet them and talk to them. While we are talking, my ancestors will tell me what the problem is. Sometimes I will just say what I think, so I have no specific way of doing it. Sometimes I will kneel down and pray to God and my ancestors and then it will just come to my mind. Sometimes I see the problem even before they come."

Traditional Healer Mirranda Javu, Western Cape, SA (2010)

In order to find out more about THs, I visited Mamgxongo, a South African TH. Like Mirranda Javu, she is a diviner—someone who sees what others cannot see. Uninitiated people cannot see because they don't know how, and don't have the spiritual gifts that allows deep insight. I visited her because of my interest in indigenous healing, not because I thought I had anything wrong with me. Lulu, a Xhosa friend of mine, accompanied me as my translator because Mamgxongo spoke Xhosa, but very little English (Figs. 6.1, 6.2).

Divination by Mamgxongo, 2009

We walked along the dusty road from Lulu's home to a modest house made of concrete blocks. Lulu had lived in Zwelihle township her entire life and described it as full of crime and drugs. Small children playing alongside the road came to inspect us, an unfamiliar group walking through their neighborhood. We entered Mamgxongo's house and sat in comfortable chairs to await her. On her wall hung her framed certificate of authentication from the '*Iinyanga Zezizwe* Traditional Doctors Organization 1998.' It was similar to the certificates of achievement that you might find in a western doctor's office.

Mamgxongo is a big woman dressed in a t-shirt with a large towel wrapped around her hips. She appeared tired and seemed uncomfortable in our presence. When her three assistants arrived, all four withdrew to another room. After about an hour they came back to the front room in full regalia.

Mamgxongo now wore a large fur headdress, as fine as a bishop's miter. She was festooned with beads—some hung in layers around her neck, and also her arms and ankles. Strips of animal hide hung from her waist making a swinging skirt. On her legs she had leather shields on which a large number of dried pupae were attached. These shook as she walked making a pleasant rattling sound. In one hand Mamgxongo held a long beaded stick, her dancing stick. In the other hand she had her whisk—the symbol of spiritual power amongst tribal leaders

(continued)

(continued)

and THs. This whisk was made from the tail of the cow that had been sacrificed when she passed her training tests and graduated as a healer.

Her assistants were now also dressed up, but less dramatically. They were healers-in-training whose job was to wait upon their teacher in every respect. The two women and a man were her apprentices. They carefully placed the tools of the healer's profession alongside Mamgxongo—an intricately beaded long pipe often enjoyed by Xhosa women, a ceremonial sword, a beaded stick that ended in a sharp steel blade.

Mamgxongo lowered herself onto the floor and sat on a small mat made from the skin of a sacrificed goat. Her apprentices now put a dried herb called *Mpepho* into an ashtray. In readiness I had brought a gift of a bottle of brandy and some money.

Lulu now indicated I should place them on the floor. Mamgxongo ignited the *Mpepho*—the herb that opens contact between people and their ancestral spirits. There was plenty of smoke and a pleasant smell wafted through the room.

I sat quietly on my chair, mystified as to what was happening. I wondered if I should also be on the floor, but Lulu gave me no instructions to do so. Mamgxongo lit the tobacco in her pipe and took three long draws. She did not look at me at all, but gazed in a different direction, with unfocused eyes. Beads of sweat appeared on her forehead. She was no longer with us, in the here and now. I believe she had entered an altered state of consciousness. After bowing repeatedly towards the floor she started incanting to her personal ancestors:

I am of the *Ncilashe* clan,
Of the *Dikela* origin,
Of *Tihopho* origin.
I worship and I pray
Let the darkness fade.
Let there be light.
I am of the *Rhadebe* clan,
The one called *Ndebele 'ntle zombini*,
With the *Mpinga*,
The *Bholokoqoshe*,
The *Maduna*.
All of you remove darkness.
Let there be light.
I ask that this disease of mine-
The disease of disliking to speak
With people of other races, the white race.
Let the disease be removed from me.
Camagu!

Mamgxongo started the divination process. She spoke in Xhosa all the time, with Lulu translating. She touched on several mild complaints that I had been experiencing. Then she surprised me by suggesting that the occasional aches in my legs were due to someone at my work who was jealous of me. I was shocked. Can physical symptoms be caused by someone else's jealousy, especially if you don't know about it?

Fig. 6.1 Traditional Healer
and assistant prepare
for our interview by burning
the herb *Mpepho* to
encourage the presence
of spirits (Photographer:
Katharine Hewson)

Fig. 6.2 Mamgxongo
shows how to mix
a concoction of roots
to encourage dreams
(Photographer: Katharine
Hewson)

Traditional Healing Extraction Treatment

Traditional healing methods involve natural products such as plants, animals, and minerals. The particular agents used for healing form a complex body of knowledge that is located within the biomes of the environment. This knowledge is mostly empirically based, but interpreted by the THs as spirit directed. Most healing approaches involve concoctions and decoctions. They are made from herbs (fresh, dried, burnt, pulverized) and animal parts (e.g., fatty tissues, organs, and skins). Sometimes dried substances are mixed with water or petroleum jelly, or simply scattered around a living space. They are usually applied as topical potions, or ingested as emetics, enemas, smoked, or worn for good luck (e.g., crocodile skin). Special dried plants may be tucked into the thatched roof of a hut to ward off lightning strikes. Healing approaches are usually non-invasive, except for a procedure called 'cutting.'

When traveling in the northern part of South Africa I met Mmabatho, a TH who lived in a small brick house in a suburb near a middle-size town. She enthusiastically invited me to observe several of her healing sessions. Mmabatho also taught me a great deal about African healing approaches. On the whole, her patients seemed to suffer physical ailments such as headaches, and stomach aches that Mmabatho diagnosed as the result of a breakdown in relationships either at home or at work. Her remedies mostly included potions for jealousy, anger, fear, and grief.

Extraction of Spiritual Intrusions by Mmabatho, 2008

On a hot summer day in the highveld region of northern South Africa I visited Mmabatho. She took me to her consulting room at the back of her house. On the walls there were shelves of numerous medicines, both herbal and animal-based in bottles of all types, old spice bottles, jam-jars, and the like. Each was labeled in clear handwriting in the vernacular of the region—Northern Sotho. On the walls were posters, a Christian prayer, the dried skin of a *meerkat* (a small indigenous animal), and a tortoise shell. Odd smells of unknown origin filled the room. I knew Mmabatho collected her own herbs from uncultivated lands, and also bought products from the *muti* shop (African medicine store) in the nearby town. She prepared her concoctions and decoctions herself.

The patient, whose English name was Jane, needed treatment for numbness, weakness and pain in both her right leg and arm. She had previously had a diagnostic divination with Mmabatho, who was now preparing the treatment for the problem.

(continued)

(continued)

Mmabatho lowered herself onto the animal skin on the floor, while Jane sat on a green chair. I was anxious about whether Jane would resent having me present, an intruder in the doctor-patient interview and treatment.

"No, no, don't worry." Jane nodded her enthusiastic agreement that I should stay.

Mmabatho prepares to throw the 'bones' to check with her ancestral spirits that her proposed treatment was correct. The spirits agreed—it was correct.

"Someone was jealous of Jane's success in her *shebeen* (a local pub that sells alcoholic drinks, such as locally brewed beer). This jealous person has sent evil spirits to her body causing the painful and weak leg and arm." Mmabatho explained the problem to me. "This type of problem is due to unnatural causes—the work of witchdoctors. These are evil people who use reverse healing to inflict harm, even death, on others."

Using a sterile razor blade she made little cuts in the patient's skin in numerous places on her arms and legs. She worked on both sides of the patient's body, starting with the well side. Then she rubbed a concoction into the cut areas. This consisted of a black mixture of four herbs that had been burned and then mixed with red Chinese petroleum jelly. Chinese petroleum jelly comes in several colors and red was important for this woman because of her gender and status in society.

Jane swore gently and shed a few tears. If an injection is painful, this treatment was rather more so—no sterile wipes, no local anesthetic. Gloves were used for rubbing in the concoction, and both Mmabatho and Jane cautioned each other that they needed to think about HIV/AIDS.

I shuddered, feeling surprise—the intrusiveness of this procedure and the minimal attention to hygiene or pain control was unexpected. As an invited guest I could only sit quietly and keep my cool. So I kept my head down and took notes.

Then Mmabatho mixed a second concoction involving the fat of a female hippo (which can be purchased in the African medicine markets) plus two dried herbs. These were for good luck—specifically to keep away the evil spirits that had initially entered Jane causing the symptoms.

"The pain is going," said Jane. A little later she declared: "Now it's nearly gone." Relief spread over her face and she wiped away her tears.

"This pain is now entering my own feet," announced Mmabatho. "It is always possible that the evil spirits removed from a patient can enter the healer. We healers must be careful about such problems." She wiped some of the good luck potion onto her own face. By the end of an hour the patient was entirely pain free, with strength returning to her arm. She paid for the treatment, and left satisfied.

Traditional Healers' Ways of Knowing

I discuss the African approach to healing in terms of their concepts of illness and healing, the consultative process they commonly use, and their approach to diagnosis, treatment, and prevention.

Understanding of illness and healing. According to traditional African IK systems, every person is a holistic integration of body, mind and spirit. In particular, the human spirit is seen as vital energy (African *num*), similar to the Chinese view of *chi*, the Indian view of *prana*, and the western view of life-force or spirit (but not the popular western views of ghosts). Africans see the entire world and the cosmos as invested with this vital energy—it is not exclusively a human phenomenon.

African medical IK conceives of illness as an imbalance or disequilibrium in the mind and body. This means, in western terms, that when people suffer from social, psychological, physical, economic, environmental or spiritual problems, they often experience psychological distress that can cause pain at many levels. Such pain is caused by disequilibrium within the body/mind/spirit complex, the presence of spiritual impurities, and/or intrusions. In order for healing to occur, equilibrium must be re-established, and the patient must be cleansed of spiritual intrusions and impurities.

African healing also involves appeasing patients' spirits, their own and those of their recently departed ancestors. The ancestral spirits might be angry with the patient for some reason—known or unknown. Spiritual interventions include rituals, the offering of a pot of homebrewed beer, vegetables such as a pumpkin, and sometimes the slaughter of animals (a chicken, goat, pig, cow) as offerings to the ancestral spirits, culminating in a feast for the community.

When patients suffer from being 'hot' (see Chap. 4), the TH cools them with wet agents such as water, and aquatic plants and animals. Specialist THs take care of birthing mothers, and others specialize as bone-setters. THs pay a great deal of attention to prevention, especially the prevention of bad luck and misfortune, both at home and at work.

Herbal medicines are widely used for prevention and also for the treatment of physical ailments. Herbs (as well as animal parts) are effective for healing numerous physical ailments, such as headaches. They are also thought to be effective for many social problems such as protecting the home, dealing with jealous people, attracting a desirable mate, becoming pregnant, and for helping dying patients and their family members.

Traditional indigenous treatments operate without the technological tools of western medicine. The introduction of western therapeutic techniques by colonial health professionals, especially X-rays, surgery, and medications (pills and injections), caused great interest in African society. They were seen as more potent healing agents for dealing with African sicknesses than traditional herbal medicines. Many contemporary Africans superimpose western healing approaches on top of their own non-western approaches, accepting the ministrations of both western healers and THs at the same time. (See Chap. 5 for a discussion of medical pluralism etc.).

THs appear to specialize in psychosocial problems, especially concerning anger, jealousy, grief, and lust. THs see illness as the result of disequilibrium within the patient's body, mind, and spirit. And they use spiritual agency to resolve the patient's disequilibrium, especially in the case of unnatural illnesses, and return the person to health—a harmonious state of equilibrium.

Consultative process. Traditional healers await their clients in a room or space set aside for healing. This room can be a hut with thatched roof or a brick and mortar room. There are decorations in the form of healing tools, skins, knives, dancing sticks, and animal skins. The healer always sits on the floor on a special mat, usually the skin of the animal that he or she slaughtered at the time of graduation. If using 'bones,' the TH contains them in a pouch, shakes the pouch well, and then throws the bones onto a grass mat in order to 'read' the diagnosis (Fig. 6.3).

The client can sit on a chair. The client may bring other people along to the session with the TH. These people may be called the therapy management group, but sometimes they are simply supportive family or friends whose presence is helpful in the process of connecting with the spirits. There is no requirement for privacy in the TH-client interaction.

First the client offers a payment 'up front' for the diagnostic process. The TH can assess the severity of the client's problem from the amount paid up front. The TH enters the diagnostic process using the 'bones.' Some THs enter an altered state of consciousness (state of non-ordinary reality) by smoking, dancing, or singing. This state is like a wakeful dream, in which the dreamer is able to control his or her cognitive actions. It can also be described as a transcendent state of awareness. The ability to enter this altered state is learned during training. In this altered state, the healer gains access to a vast amount of information and finds answers to questions, such as diagnosis (Harner 1990).

Fig. 6.3 Mmabatho
prepares to throw the bones
(Photographer: Peter Hewson)

Next, the TH will state the diagnosis and make therapeutic recommendations. Then the client will request the medicines (*muti*), the treatment, and also the preventive strategies to promote better health. There is a second payment at this point that combines the cost of the medications and the client's satisfaction with the TH.

Diagnosis. In African medical IK, THs use spiritual powers and non-ordinary reality to 'see' causes of illness. THs confer with their own ancestral spirits as well as those of their clients in order to diagnose problems (Hewson 2011).

A common diagnostic practice amongst southern African THs is to 'throw the bones.' The 'bones' are a collection of several objects collected by the TH e.g., small animal vertebrae, as well as shells, stones, or cultural objects such as dice, coins, bullets, and beads. The TH attributes each object with a particular meaning e.g., happiness or illness, that provides information for the diagnostic process. The TH throws the 'bones' onto the mat, at the same time invoking the ancestral spirits with a prayer. The TH then reads the 'bones' for their significance in terms of how they fall in relation to each other.

Not all THs use the 'bones' for diagnosis. Some THs prefer to smoke a ceremonial pipe, while others dance and sing. Still others simply pray to their ancestors. Many THs use a combination of approaches. Their intention is to enter an altered state of reality and to access their ancestral spirits. In this state they receive answers to their questions about their patients.

When making a diagnosis a TH looks at the patient both physically and energetically—seeing the physical body in connection with the energetic forces that surround the body. This is referred to as divination, which may be through dreams, or through entering into a non-ordinary reality—an altered state of consciousness (Harner 1990; Katz 1982). People enter this state in various ways such as drumming, dancing, smoking particular herbs, prayer, or a combination of methods. Being in non-ordinary reality is similar to wakeful dreaming, where everything can be remembered, unlike hallucinating or being in a trance (Harner 1990).

The TH needs an open and intuitive mind in order to access his or her natural wisdom and compassion. Being in a state of non-ordinary reality involves visualization or imagination and may be similar to self-hypnosis or deep meditation. Whatever it is, THs believe that when in this state of non-ordinary reality their particular ancestral spirit (mother, grandfather, aunt, etc.) mediates for them and speaks to them on behalf of the patient. The spirit provides insight into the spiritual and physical needs of the patient.

THs do not have access to modern tools such as X-rays, MRIs or CAT scans to verify their findings. They operate with the inner confidence provided by the spiritual voices of their particular ancestors. When THs are out of their depth, they often refer their patients to another TH, or to western medical practitioners, e.g., for HIV/AIDS.

THs believe that their spirits will inform them about the cause of a patient's illness. Most of the THs with whom I talked said that their dreams provide the answers to their questions—the ancestral spirits speak through their dreams. One of them told me that the answers simply come to mind. Another mentioned that it is like hearing a little voice in your ear.

Treatment. THs rely heavily on the treatment 'advice' they receive from their ancestral spirits. This advice may come in the form of a dream, an intuition, or a physical sign of some sort. For example, after divination (throwing the bones) Mmabatho had received advice for the appropriate treatment of her patient. She repeated the divination process when Jane returned for treatment to check the accuracy of the original advice.

For unnatural causes, THs purify the patient's body by removing impurities, especially evil spirits, through the use of purgatives, emetics, and cutting. Generally such treatments include the ingestion of herbs, animal parts, and minerals. In the story involving Mmabatho the herbal remedy was introduced through the skin of the patient.

A TH may not provide the same treatment for two patients suffering from the same complaint. The TH is likely to view the similar presentations of illness as two different problems depending on the contexts of the patients. For example a man with a headache could be treated differently to a woman with a similar complaint. Contextual factors, such as a patient's gender, age, life events give rise to different treatments. A patient's spiritual status, e.g., harmony (or lack of it) concerning ancestral spirits, and psychological status, e.g., feelings of anger, jealousy, fear, etc., are taken very seriously. THs do not seem to generalize or standardize treatments as happens in western medical practice (e.g., standards of care). African medicine appears to specialize in existential, interpersonal and spiritual suffering.

Prevention. A large component of indigenous African medicine concerns prevention. It is very important for people to ward off negative spirits. An example of a bad or harmful spirit is the *tokoloshe*—a mischievous or evil spiritual being that is responsible for many of life's misadventures and accidents. Curses from living people, and the ill will of angry ancestral spirits are a prevailing potential problem for Africans. These negativities must be prevented whenever possible. The preventive techniques (like cures) involve herbal potions, creams, and ritualistic actions such as sacrificial offerings to the ancestors. This version of preventive medicine differs markedly from that of the west.

Comparison of Western and Traditional Healing Procedures

Using the same categories as above, I conceptualize some of the differences between the two approaches to healing:

Understanding of illness and health. Western medicine treats the biological causes of disease—the biomedical scientific approach. Traditional African medicine is concerned with both natural causes of illness (infections, malfunctioning of the body, etc.) and spiritual cases. But there is little place in western science for what is perceived as magical, fantasy or faith-based claims—the world of dreams and spirituality is excluded. It is no longer acceptable for a western doctor to make medical decisions on the basis of a hunch or intuition. Western researchers have documented that patients hide their psychosocial and emotional concerns because of social stigma (such as mental illness), and may actively resist psychiatric or psychosocial diagnoses, preferring a concrete physical diagnosis.

Diagnostic procedure. The western doctor first engages with the patient in a private space and creates a relaxed, communicative atmosphere. The doctor then obtains the 'history of the present illness,' as well as 'past medical history' and factors that might be relevant to the case (such as age, living arrangements, work related information, marital status, number of children). Then the doctor performs a physical examination and medical tests may be ordered. The doctor makes a diagnosis and tells the diagnosis, the proposed treatment and plans for follow-up. He may also tell the patient the cause of the illness (etiology) and future prospects (prognosis) for getting well.

African THs do not appear to do a physical examination or empirical tests. They rely heavily on the information given to them by their spirits. There is little empirical verification of appropriateness or efficacy of the diagnoses or treatments. If a treatment does not work, the TH will try again. If it still does not work, the TH will send the patient to a different TH (a referral), or to a western doctor.

Western doctors use technological and scientific tools to recognize the pathology or patho-physiology in their patients. Using clinical reasoning and evidence-based medicine provides a measure of confidence that their diagnoses are accurate and dependable.

The confidence of western physicians is based on a large fund of knowledge, meticulous logic, management of sterile environments, high technology, appropriate medical protocols, and ongoing research. The confidence of THs is mainly based on information provided by spiritual beings, accompanied by their extensive knowledge of natural herbal and animal remedies. The healing power of African THs is provocative and difficult for western scientists to accept. If a diagnosis and treatment plan is not based on reliable science (such as experiments and clinical trials), it is seen as quackery.

Treatments. The American Food and Drug Administration (FDA) makes and enforces rules for the protection of society. These concern the safety of therapeutic treatments and pharmaceutical drugs, and also of industrial and agricultural food products. There is little oversight of African medical approaches.

Prevention. The increasing importance of preventive medicine in western medicine includes vaccinations, inoculations, and a focus on healthy life styles. African THs pay a great deal of attention to prevention, mostly in the form of warding off bad spirits and preventing the malevolence of evil-doers in the form of witchcraft. Just as spirits can work benevolently for the good of humans, they can work malevolently. Witchcraft in African life is focused on the notion that misfortune is caused by other people or spiritual beings that may be jealous, angry, or vengeful.

Witchcraft in African Medicine

Witchcraft has helped Africans explain the vicissitudes of their lives, especially the disparities in the distribution of good fortune and material assets to people within a social group. Differences in the levels of success, wealth, health and illness are not understood in terms of statistical probabilities, as in western societies. Such inequalities give rise to anger, jealousy, greed, grief, and numerous other affective and emotional states. Witchcraft is often presumed to be the cause of such misfortunes. Just as health and illness are the obverse of each other, so are healing and witchcraft. Healing and witchcraft can be viewed as metaphors for success *versus* failure, or good-luck *versus* bad-luck. Ashforth (2005) describes the concept of witchcraft as a function of both spiritual and material insecurity of Africans.

Much African traditional healing functions as a consequence of, and in reaction to witchcraft as a way of making sense of adversity. While almost no healers confess to practicing the dark arts of witchcraft, the writings of contemporary anthropologists (Ashforth 2005; Janzen 1978; West 2005) suggest that witchcraft is a central parameter in their work. This is because in general, illness is considered to be either God-made or man-made. God-made problems occur naturally and include most illnesses such as infections and broken bones. On the other hand, man-made illnesses are caused by individuals using witchcraft to get back at their enemies. Specialists in witchcraft are called witchdoctors. They are capable of causing psychic misadventures such as spiritual intrusions in another person (Ashforth 2005). A curse by an enemy has destructive power and can cause illness and even death in the recipient. Anger from ancestral spirits who have been neglected or disobeyed can cause similar devastation. If, and when a TH divines witchcraft as the cause of an illness, the TH must fight against it, and must protect clients from future witchcraft attacks.

Early colonists noted that African people live in constant fear of evil spirits and witchcraft (Vaughan 1991). This is true to the current day in southern Africa, where fear of witchcraft is still a factor driving the social, cultural, and religious intellectual system of Africans (Ashforth 2005).

We Can Teach the Children: Additional material to this book can be downloaded from http://extras.springer.com

References

Ashforth, A. (2005). *Witchcraft, violence, and democracy in South Africa*. Chicago: University of Chicago Press.

Eliade, M. (1964). *Shamanism: Archaic techniques of ecstasy*. Princeton: Princeton University Press.

Harner, M. (1990). *The way of the shaman* (3rd ed.). New York: Harper & Row Paperback.

Hewson, M. G. (1998). Traditional healers in Southern Africa. *Annals of Internal Medicine, 128*, 1029–1034.

Hewson, M. G. (2011). Indigenous knowledge systems: Southern African healing. In M. Micozzi (Ed.), *Fundamentals of complementary and integrative medicine* (4th ed., pp. 522–530). St. Louis, MO: Elsevier.

Hewson, M. G., & Javu, M. T. (2011). Indigenous knowledge concerning health, illness and medical education in Southern Africa. *Medical Education, 25*(2), 13–18.

Janzen, J. M. (1978). *The quest for therapy: Medical pluralism in Lower Zaire*. Berkeley: University of California Press.

Katz, R. (1982). *Boiling energy: Community healing among the Kalahari Kung*. St Louis: The President and Fellows of Harvard College.

Mbiti, J. S. (1985). *African religions and philosophy*. London: Heineman Educational Books.

Toulmin, S. (1972). *Human understanding* (The collective use and evolution of concepts, Vol. 1). Princeton: Princeton University Press.

Vaughan, M. (1991). *Curing their ills: Colonial power and African illness*. California: Stanford University Press.

West, H. G. (2005). *Kupilikula: governance and the invisible realm in Mozambique*. Chicago: University of Chicago Press.

Chapter 7
Education of Traditional Healers

Abstract I describe how individuals decide to become traditional healers, their initiation process, training, accreditation, and ongoing education. A traditional healer describes how she was 'called' to be a healer. I describe some differences between becoming a healer in Africa and the west.

Understanding indigenous training approaches, especially of traditional healers, provides insight into the nature of the traditional healers' (THs) professional knowledge as well as its epistemology. Cassidy (2011) outlined the social and cultural factors attributed to medical professionalism: the decision to become a professional; the educational process; the accreditation of practitioners; and the professional organization of practitioners that monitors and maintains standards. For each factor I compare the indigenous approach with the western equivalent.

The Decision to Become a Healer

In traditional African healing systems a person is 'called' to become a healer by the ancestral spirits (Hewson 2011; Bührmann 1984). The call is perceived through two mechanisms. First, the person experiences a mysterious illness that is unresponsive to normal treatments. This can be blindness, headaches, lameness, depressive symptoms, unremitting misfortune, or a wide range of other complaints. Second, the person has unusual dreams that seem startlingly significant. This will cause the person to seek advice from a TH as to the meaning of the dreams. The healer may determine that the person is suffering from a spiritual sickness and experiencing a call to become a healer. The person will check out this possibility with his or her elders, and decide whether or not to enter training. Some people who are 'called' to become healers resist the calling. This is considered dangerous because the ancestral spirits can punish the person by increasing the spiritual sickness with the possibility of death.

Although the TH will become a person with social power, the life of a TH is generally considered onerous and disruptive of normal living. It is seen as dangerous (because when dealing with powerful spirits, one can be caught in spiritual power plays and be attacked oneself), challenging (because it is not always easy to resolve people's problems concerning health or misfortune), and dirty (because the healer has to find and prepare the therapeutic herbs and animal parts, and deal with sickness and death).

Mirranda Javu's Calling

"I had this dream in 1992 and I got sick. I was not denying it, but I couldn't understand it because I grew up as a Christian. We used to attend church every Sunday and I enjoyed that. So the church people were shocked. But when I had that dream I explained everything that was going on to my family.

In my dreams many things were unusual and could not be explained. But they were dreams that were indicating that I have to accept the call. So I went to a traditional healer in my community for divining, just for confirmation. I was the first one in my family. The traditional healer confirmed that I was having this calling to become a healer myself."

Interview with Mirranda Javu (2010)

Xhosa healers in South Africa have an initiation process that involves several stages in order for the candidate 'to become what he must become' (Bührmann 1984, p. 71). In the west, individuals decide whether to engage in a medical career on the basis of a personal attraction to the field, academic counseling, or family support. Based on standardized test scores, essays, and interviews with medical school personnel, the potential physician engages in a 7-year (or more) training program.

I interviewed a senior medical educator about her reactions to Mirranda Javu's statements. Janine C. Edwards Ph.D. was most recently Professor and Chair of Medical Humanities and Social Sciences in the College of Medicine at Florida State University, Tallahassee.

African and Western Motivations to Become Healers

"The concept of a calling is somewhat surprising to me. It reminds me of the Hebrew prophet Jeremiah, who, when called by God to be a prophet, replied, 'Why me, Lord? I do not want to do this.' The deep connection with spirits has been so denigrated in western society that most westerners find it incomprehensible that spirits could call a person to do something. Our society is so focused on empirical evidence that we reject the concept of spiritual connections. Rare is the physician who feels that he or she is 'called' to the vocation of practicing medicine.

Some western young people are altruistic—that is, they sincerely want to care for sick people. These westerners make the best physicians because they are eager to learn all the science and clinical medicine they can so that they will

(continued)

(continued)
be prepared to take care of people who come to them. Some people are motivated by their experience of serious medical conditions in themselves or in family members. These experiences provide a lifelong motivation to help others.

In the western tradition the conventional meaning of a professional is a person who possesses a specialized body of knowledge that is used for the benefit of others. A few people have a genuine thirst for knowledge of science and medicine that compels them to learn as much as possible and to continue learning throughout their lives. These are the real scientists among physicians.

Other young people become interested in medicine because they know that they will make a lot of money practicing medicine. It is quite possible in the United States for a physician to become a highly skilled subspecialist earning half a million dollars a year or more. Insurance companies, pharmaceutical companies, and financiers have made the profit motive in healthcare highly visible. Delivery of health care is widely considered a business, which, of course, means that the profit motive is primary. This is a corruption of the traditional professional value of physicians—to be of service to humankind.

Still other westerners are motivated by the high level of prestige accorded to physicians in our society. In most cities and towns in the United States, medical doctors are considered representatives of the highest levels of society. They are invited to be honorary members of societies, and to raise money for all types of medical causes. They also belong to elite clubs and become leaders in social activities, such as balls and parties, and golf tournaments. The title 'Doctor' carries with it respect and deference in every situation. Doctors can obtain large amounts of financial credit, which enables them to engage in business deals that make them even more money. Finally, some young people become physicians because of their parent's 'legacy' motive—being a physician may be a tradition in the family. This has created many unhappy, but apparently successful physicians."

Interview with Janine C. Edwards, Ph.D., medical educator

Initiation Ceremony

According to Vera Bürhmann, after the initial dreams and discernment process involving the candidate's family and elders, the master healer organizes the first river ceremony. This involves brewing a large quantity of special beer. While the beer is brewing (a 3–4 day process) the candidate remains in isolation while the family and friends engage in rhythmic drumming, clapping, singing and dancing, usually inside a hut. Then in the early light of morning a small group with painted faces or wearing masks walks in silence to naturally flowing water, or the sea. They make offerings to the 'River People' who inhabit the waters. These offerings are gifts in the form of food products, beads, or other valued objects. The group watches the water for evidence that the River People have accepted the gifts. The small group returns in silence to the household and reports on the response of the River People to the acceptability of the candidate for training. If it is a positive report, all the assembled people relax and rejoice. The beer is consumed, and the early morning jubilation starts the serious training process for the candidate.

On a recent trip to Zimbabwe in 2007, I participated in an initiation process, which I describe in more detail in Chap. 11. The process followed similar steps as those described in Chap. 2 (my first meeting with a TH in Lesotho), and also for the Xhosa people, as described by Vera Bührmann (1984).

Training Process

After being 'called' and being accepted for training in the African healing knowledge and skills, candidates embark on a rigorous, prolonged, often expensive training with a master healer. The process is not strictly regulated. The emphasis is on subjective witness of healing efficacy because dreams of both the trainees and trainer drive the process. The ancestral spirits guide all decisions, such as when a person decides to enter training, and with whom to be apprenticed. The spiritual entities guide the trainee's curriculum, promotion through the various stages of the training process, and confirm when the process is completed. The guidance comes in the form of unusual dreams, which are interpreted as signs and messages from the ancestral spirits.

The indigenous curriculum tends to be idiosyncratic and depends on numerous factors concerning the trainee, trainer, and the environment. There are, however, generally accepted stages that must be accomplished: the initiation process, demonstration of healing skills, ritual steps to increasingly complex skills, final tests, and graduation.

The indigenous training process is of variable duration, and relatively speaking, it costs a lot. The trainee must pay for board at the trainer's compound, a down payment to the trainer, and a final payment at graduation. The trainee must procure the various elements for healing, specific clothing and ornamentation as specified by the ancestral spirits (e.g., beads), and items needed for important rituals. For example, at graduation the trainee supplies an animal (chicken, sheep, goat, or cow) for ceremonial slaughter. A community feast follows this event involving the cooked meat, *putu* (corn porridge), vegetables, and generous supplies of home-brewed beer.

Mirranda Javu on Her Training

"You have to have a trainer because even though your dreams are leading you, you can't be doing everything by yourself. Your ancestors will tell you things in your dreams, but you have to get someone to assist you and to interpret your dreams as well.

In the training process we play a game that uses the skills needed in *unontongwana* (finding lost people or things). The trainer hides something, and as a trainee you learn to look for it, find out exactly where it is, what it

(continued)

(continued)

is, what it looks like, and be able to go and get it. It is not easy—it's a tough game! This game is a form of training that helps trainees in finding stolen or missing things.

As a trainee I was curious to know everything because I wanted to check which skills were in me. I used to be the first to consult our clients. This has helped me a lot because if you are not interested in learning you take a long time to know things. So you have to push yourself to learn.

During training there are also things that you are instructed not to do, or food you may not eat. You do not have a right to ask why, because these (prohibitions) may help you through your training processes. So I was very careful."

Interview with Mirranda Javu (2012)

The interpersonal relationship between western medical teachers and their students persists, but does not seem to be upheld as fervently as in the indigenous situation. Strong bonds between western medical teachers and their students do develop and may last a lifetime, but they are not expected. Professional mentoring has received recent attention in medical training as well as within the professional arena (Murray 1991). Western medical trainees tend to develop ongoing relationships with their professional colleagues, especially around medical research projects and professional activities at the national level.

Western medical training is highly regulated, although it was not so regulated before the era of modern medicine. National bodies provide specifications for medical curricula at medical schools and residency programs, and the final Boards exams of the various disciplines ensure that core competencies are maintained. Medical training in the USA is arduous, prolonged, and expensive. While successful candidates can generally be assured of good financial compensations, the life of physicians is generally demanding.

Western Medical Training – Interview

"There are some similarities between Mirranda's description of her training and training in the west. Competition to obtain a place in a medical school is keen. Young people study hard, work with coaches, and take classes to prepare for rigorous entrance tests. They also spend significant amounts of money to be selected by medical schools.

The training of western physicians is highly technical and, like the training of traditional healers, is very difficult. It involves many years of apprenticeship, working in teams of trainees with one or two experienced physicians. Hardships, such as long hours of work, and unpleasant working conditions are

(continued)

(continued)
an essential part of western medical training. The tests for western medical trainees involve a large number of written exams during the four years of medical school as well as national tests for licensure in order to be able to practice medicine."
Interview with Janine C. Edwards, Ph.D., medical educator

Accreditation

Accreditation in a TH apprenticeship is based on the approval of the master healer. This is done in conjunction with the dreams of both the trainee and trainer, in which the ancestral spirits indicate whether the trainee is ready for independent practice (Hewson 1998). Approval is also obtained from the patients treated by the apprentice, together with community support. This approval is often made manifest with gifts of items needed by the trainee. The more patients and community members show their support for a trainee, the sooner he/she graduates.

Trainees are also tested. These tests might include the ability to recognize and use herbal and animal medicines, knowledge of appropriate dosages, or skills such as finding a hidden object. Most important, the trainee has to show spiritual connectedness – the ability to 'see' the cause of problems (diagnosis) and find solutions to problems through the mediation of ancestor spirits.

When trainees are ready to graduate, they receive a dream about a particular animal, such as a perfectly white goat. This dream indicates that the trainee needs to locate a particular animal that will be ceremonially slaughtered at the graduation ceremony. The trainee will take certain parts to be used in future healing ceremonies. For example, the skin will be cured to become the mat upon which the healer sits when doing healing, and the tail will become the whisk, the symbol of African THs.

At the end of the training the healers are acknowledged as people with enhanced knowledge and skills that set them apart from common people. The rigor and hardships of the training process appear to be an important component of the resulting status and power of healers.

Mirranda Javu on Accreditation
"The trainer won't know that you have to move if you do not dream about it. First you have to be guided by your ancestor then the teacher will interpret the dream and take you to another phase, or to graduate.

As a trainee you will show signs that you are ready to move to the next step. Sometimes you don't dream explicitly that your ancestor is telling you something, for example, you may dream about beads with certain colors, or certain animals. The trainer may interpret these dreams as signs that you can

(continued)

> (continued)
>
> go to the next phase, or to graduate. For example, you will be allowed to stand up in front of the trainer instead of always kneeling down to show respect. So the teacher must be good and able to interpret dreams. Eventually the trainer tells you 'OK, you have to move to the next phase now.' But anyway, the trainer knows that, when the day comes, there will be proof in the trainee's dreams.
>
> So when you are trained, the trainers observe each trainee, noting who is capable of doing what. For example, one trainee may be good at preparing herbal treatments, and the other one is good at divining or perhaps in *unontongwana*."
>
> Interview with Mirranda Javu (2012)

Vera Bührmann, in her account of Xhosa TH training, tells that at some point close to the finalizing of the training process, there is a second ceremony. This time the celebration involves the slaughter of an animal (preferably a pure white goat). The animal parts are 'ritually incorporated' by the trainee (Bührmann 1984, p. 77). The meat is cooked and shared amongst the trainee and his or her relatives and friends. The point of the celebration is to ensure the presence and participation of the ancestral spirits, and invest the trainee with symbols of the healing profession such as the white skin of the goat.

All the THs in my several investigations went through initiation and graduation ceremonies similar to those described here. In Chap. 11, I describe my visit to Zimbabwe to study with a TH. During this visit one of his students, a Shona woman, graduated. Her ceremony involved extensive preparation of food (meat and corn) for the whole group involving her family, friends and visitors. She brewed indigenous beer made from sorghum. Persistent, loud drumming and singing evolved into a trance dance invoking states of altered consciousness in most people in the group. During the trance dance the new TH demonstrated her powers of divination by proclaiming 'truths' about selected people. The process was preternatural, and after a while I was happy to extricate myself from the group and withdraw to bed.

Western medical trainees are accredited on the basis of objective test scores in national certifying/licensing exams e.g., National Boards. While there is usually no patient input, the recent trend in medical education involves using standardized patients to test medical students' diagnostic, clinical, and communication skills. Western medical training is rigorous, prolonged, and expensive, with an emphasis on objective measures of competence designed by professional boards.

Continuing Education

South African THs are required to register with, and belong to a National Association of Traditional Healers. They meet frequently and the meetings serve as opportunities to share cases and discuss care options. They also meet to discuss the practices of other THs with concern about appropriate and inappropriate healing strategies. The 'calls' of new apprentices are also topics for discussion. At these events THs wear their full regalia (head dresses, extensive beads, animal skins), and take their whisks, dancing sticks, and other accessories of their profession to the meetings. They share a meal, and there is dancing, singing and music-making, especially drumming. The healers may also engage in communal spiritual 'trance-dances' during which they connect with their ancestors.

THs need to have strong professional relationships because there is competition amongst them for access to therapeutic herbs from the countryside. Some THs travel great distances to find the necessary herbs and animals (e.g., snakes, scorpions). If they can't find them they must pay for them in *muti* (medicine) markets. Professional jealousies are possible, and everyone fears the powers of witchcraft (interpersonal and spiritual malevolence).

The relationship between a trainer and the TH trainee tends to be life-long. Many THs return regularly to their trainer's compound for an update. These trips may involve large distances and considerable expense. This relationship is honored as much as (or even more than) that of a child–parent relationship.

Mirranda Javu on Her Life as a Traditional Healer

"So I went through all those trainings successfully and now I am an independent traditional healer and also a trainer of others. I've had eighteen years in this field. I also have eight trainees who have finished and graduated, and thirteen who are still trainees.

Both my trainees and clients call me and make appointments to see me because I am busy. They are also busy. I always make time for my clients even if it is late in the evening, early mornings, or weekends. As a healer you must be available for your clients and for your trainees as well."

Interview with Mirranda Javu (2012)

Both western and traditional healers engage in regular and frequent meetings with other healers. Western medical professionals are required to engage in continuing medical education and to generate a specified number of continuing medical education credits (CME). The requirement keeps doctors' knowledge and skills up-to-date in a rapidly changing field of medical research. Conferences are favored but expensive. Innovative approaches to CME involve online courses as well as journal-based interactive programs.

Traditional healers hold meetings involving traditional ceremonies (dancing, singing, etc.) and deliberations on topics including the politics and economics of traditional healing, maintenance of consistent, spirit-directed practices and avoidance of false practitioners. With the encouragement of the South African government, there are increasing numbers of TH organizations. For example, Mirranda Javu has a leadership role in the *Indibano* Traditional Healers Association in the Western Cape, South Africa, and serves in the national organization of THs in South

Africa as well as several national committees concerning traditional healing in South Africa.

These comparisons between the training processes of THs and western medical personnel show that, despite the differences, there are many points of commonality between western medicine and traditional African healing, and the respective approaches to training. While western medicine has immense scientific and technical knowledge and skills at its disposal, it lacks one component in which traditional healing excels—an understanding of the holistic workings of the human body, and the connection of the body with mind and spirit.

> **Learning Trust and Respect in Medical Encounters**
>
> *Mirranda Javu.* "Throughout the training to be a healer you learn many things such as how to treat people with respect, how to talk with people. I respect myself and people respect me and I also respect them. I respect my ancestors as well."
>
> *Janine C. Edwards.* "We westerners need to trust in our spiritual nature, and to connect with forces beyond our rational knowing. We can learn from the trust that traditional African patients place in healers' powers of healing. The western physicians' contract with society is currently broken because society can no longer trust that physicians place the patients' interest above their own self-interest. Physicians need to re-establish themselves as trustworthy. Western patients need to learn again to place trust in their physicians."

Trust and respect are essential components of a functional medical relationship. A patient who doubts the doctor's knowledge and skills for any reason is unlikely to follow the doctor's medical advice and prescriptions. According to the Institute of Medicine nearly 50 % of patients in the USA—about 90 million—have trouble understanding their doctor's recommendations (Nielsen-Bohlman et al. 2004). This is a large number of people, many of whom are unlikely to comply with the doctors' recommendations. This has serious negative consequences for the country's health care.

People often do not understand their doctors. The reasons are many—the patients may be too old or too young, be cognitively impaired, be immigrants with language difficulties, or have different cultural beliefs. The problems with language difficulties and different cultural beliefs are of particular interest because they are so prevalent in the modern world as well as within individual nations. It is especially problematic when western doctors minister to patients from other cultural groups. Apart from using techniques such as speaking slowly, writing take-home notes for the patient, and using drawings, the basic need is for doctors to respect their patients, and consequently for patients to be able to trust their doctors.

Being respectful could manifest in doctors paying careful attention to their communications with their patients, both cognitively and affectively. This involves western doctors: taking time to understand and apologizing for their lack of understanding; being completely open to what patients tell them, especially concerning cultural practices and beliefs—even if it seems outrageous; being empathetic with suffering patients whatever the cause; and building supportive ongoing relationships with their patients.

Ongoing research in the western world of medicine is foregrounding the doctor-patient relationship as an essential component in good medical care (Kern et al. 2005). And most medical schools and medical training centers now include mandatory coursework in this area. Moreover, many medical training centers now also include courses and even fellowships in integrative medicine—knowledge about, and integrating complementary and alternative medicine into western medicine and medical training programs (Hewson et al. 2006).

All healing systems seek to alleviate suffering and promote health and wellbeing. When working across the cultural divide there are opportunities for mutual learning. African THs have much to learn from western medicine in terms of anatomy of the human body, pathology, pathophysiology, medical ethics, modern approaches to testing and treating patients, and so on. Western doctors can learn a great deal about the body-mind complex and the important role of spirit in the business of healing, curing and caring for patients.

We Can Teach the Children: Additional material to this book can be downloaded from http:// extras.springer.com

References

Bührmann, M. V. (1984). *Living in two worlds*. Cape Town: Human & Rousseau.

Cassidy, C. M. (2011). Social and cultural factors in medicine. In M. Micozzi (Ed.), *Fundamentals of complementary and integrative medicine* (4th ed.). St. Louis, MO: Saunders Elsevier.

Hewson, M. G. (1998). Traditional healers in Southern Africa. *Annals of Internal Medicine, 128*, 1029–1034.

Hewson, M. G. (2011). Indigenous knowledge systems: Southern African healing. In M. Micozzi (Ed.), *Fundamentals of complementary and integrative medicine* (4th ed.). St. Louis, MO: Saunders Elsevier.

Hewson, M. G., Copeland, H. L., Mascha, E., Arrigain, S., Topol, E., & Fox, J. (2006). Integrative medicine: Implementation and evaluation of a professional development program using experiential learning and conceptual change teaching approaches. *Patient Education and Counseling, 62*, 5–12.

Kern, D. E., Branch, W. T., Jackson, J. L., Brady, D. W., Feldman, M. D., Levinson, W., & Lipkin, M. (2005). Teaching the psychosocial aspects of care in the clinic setting: Practical recommendations. *Academic Medicine, 80*(1), 8–20.

Murray, M. (1991). *Beyond the myths and magic of mentoring*. San Francisco: Jossey-Bass Publishers, Inc.

Nielsen-Bohlman, L., Panzer, A. M., & Kindig, D. A. (Eds.). (2004). *Health literacy: A prescription to end the confusion*. Washington, DC: National Academies Press.

Part IV
Research in, and Applications to, Science and Medical Teaching

Chapter 8
Educational Research on Indigenous Knowledge in Science Teaching

Abstract In this chapter I return to thinking about teaching science. A failed university course in Lesotho suggested a reexamination of basic principles of teaching science in Africa. I offer a model of teaching and learning, and discuss the criteria for a functional educational curriculum. My recent investigations of the indigenous knowledge of traditional healers concerning the education of children in both South Africa and Lesotho indicate agreement amongst the traditional healers. They also suggest a way forward in trans-cultural teaching.

The ideas of indigenous people about science, and how to teach it became central to my research in the years 2007–2012. In this research I brought together various aspects of my career: theories of teaching and learning; science and science teaching; indigenous knowledge, traditional healers (THs) as holders of indigenous knowledge (IK); and suggestions for improving teaching in order to improve the learning of science.

When I was teaching science education at the University of Botswana, Lesotho, and Swaziland (UBLS) in the 1960s, the science faculty reached an impasse. Somehow, our students in the first year of their studies were doing very badly in all the science disciplines: chemistry, physics and biology. The respective faculty met to decide what to do about it, because we all knew that the students were not stupid.

> **'Foundations of Western Science' Course, 1968**
>
> "It seems like something is missing," said the chemistry lecturer. "I teach on about atoms and molecules, but nothing seems to stick. One of my students once said 'Sir, if you can show me an atom, put it in my hand, maybe I can learn it,' and he held out his hand as if to receive an atom or two."
>
> "My students don't seem to like the botany I teach them. It is so strange because I love knowing about plants, their structure and their names. I also teach about local plants. But they don't get excited by anything I teach," complained the botany lecturer.

(continued)

© Springer Science+Business Media Dordrecht 2015
M.G. Hewson, *Embracing Indigenous Knowledge in Science and Medical Teaching*,
Cultural Studies of Science Education 10, DOI 10.1007/978-94-017-9300-1_8

(continued)

The physics lecturer just shook his head. "No, I can't get my students to enjoy physics and when we get to the calculations, it is hopeless."

Someone suggested that the students lacked the historical perspectives of western science. Another thought that students' ignorance about philosophy of science might be the problem.

"I think they need to know and understand the origins of western scientific thinking. Let's offer a mandatory course for all science students: Foundations in Science."

The faculty put together a 3-month course with students meeting once a week, and a pre- and post-test. Our topics included the history, philosophy and sociology of science. The history of science component introduced pre-Socratic scholars' notions about the universe, Greek science and the controversies surrounding the nature of matter, Ptolemy's geo-centric system and Copernicus' heliocentric system, the mechanics of Aristotle, Newton's physics, Descartes' separation of the mechanical world from the science of the mind, Einstein's and Bohr's arguments about the nature of reality, etc. The philosophy of science component included forms of reasoning in science (empiricism and rationalism), the nature of theories and models in scientific explanations, the nature of causality, probability and determinism, etc.

The faculty formed a teaching team and diligently offered the lectures. We assembled a student handbook of relevant readings, and arranged occasional student discussions with a facilitator. In fact, as scientists, we thoroughly enjoyed the topics that formed a refresher of our own scientific antecedents.

The Foundations of Science course started well but by the end of semester the dropout rate was disconcerting. The scores on the pre-test were predictable (mostly in the region of 20–30 %). The post-test scores were about the same—there was little change. After all the hard work the faculty were mystified and depressed.

Exploring the Failure of the 'Foundations of Science' Course

The Foundations of Science course is an example of a failed course. Even though the lecturing team believed they had developed a good course, and were enthusiastic about it, they failed to hook the first year science students' interest and little was accomplished.

In the following section I describe a model of teaching and learning as a means of interpreting the failed lecture course.

Model of Teaching and Learning

This model has three main components: the goals and objectives of the curriculum, teachers' teaching strategies and students' learning strategies, and outcomes (test and/or examination scores). The planned goals and objectives (such as the teaching topics for the program, course and lessons), should lead to the desired outcomes (the measured gains in knowledge, attitudes and skills). The intentions of the teachers to teach in a particular way need to be consistent with the intentions and abilities of the students to learn a topic for a particular purpose. The teacher's selected teaching strategies need to be compatible with the preferred learning strategies of the students. The success of the teacher is measurable by the success of the students in achieving the specified goals and objectives. The success of the whole curriculum rests on the consistency (congruence) between the components of the curriculum.

Teaching and learning model

Component	Teacher	Student
Goals and Objectives	Knows the topic involving specific knowledge, attitudes and skills	Wants to know the topic, and has prerequisite knowledge, attitudes and skills
Strategies	Selects and uses appropriate teaching strategies	Prefers some learning strategies but not others
Outcomes	Succeeds or fails based on student success or failure	Succeeds in, or fails examinations and tests

Adapted from Hewson and Hewson (1988)

Goals and objectives. These consist of the specified knowledge about a particular topic, attitudes towards that topic, and relevant skills (both cognitive and physical) that are the focus of the teaching event. A curriculum exists in the context of the school and its environment, its historical, religious, and cultural setting, as well as the dominant language of the students and their socio-economic status. The specified curriculum goals and objectives can include knowledge about scientific facts, the nature of science, the philosophical basis of this knowledge, and the methods of investigating it. It includes scientific attitudes such as truth-telling and scientific rigor. It also includes the skills of science, such as reasoning skills, problem-solving skills, experimental skills such as setting up controlled experiments and manipulating variables, managing scientific apparatus, etc.

Outcomes are assessed in local or national examinations. A successful curriculum is one in which the proposed goals to teach a particular topic result in appropriately measured outcomes—students succeed in passing the examinations set by the teachers (or national bodies). Poor student performance in examinations suggests a failure of the entire curriculum. There are many factors for this failure such as problems concerning the students, the setting, or the teachers themselves.

Teachers and teaching strategies. Teachers need knowledge of the curriculum topic and appropriate training to teach the knowledge, attitudes and skills. Most teachers undergo teacher training to qualify them to teach to specific levels. However, teachers have experienced the teaching of others (as students). These experiences often preset their own expectations of how teaching should take place, and what counts as good teaching. Most teachers have preferred instructional strategies. For example, some teachers like to use Socratic questioning, while others prefer giving the class an initial quiz followed by an exposition on the topic. Many teachers still prefer using worksheets that the students complete, or simply lecturing.

Students and learning strategies. Students need to have the motivation to learn the curriculum. They need the prerequisite knowledge, attitudes and skills concerning the curriculum theme and topics. Students often have prior knowledge that comes from their cultural and social backgrounds, but it may differ significantly from the knowledge, attitudes and skills assumed and required in the curriculum. This can serve as a barrier to their further learning (Hewson and Hewson 2003). It is helpful to the students when the teacher knows their prior knowledge and can integrate it into the lesson plan. Students also have preferred learning modes, for example, some prefer to learn by doing (such as projects or laboratory work), others like participating in group discussions, and yet others prefer quiet study such as reading, and doing problem-solving exercises.

A Functional Curriculum

All six components of the model need to be congruent—the teacher and students need to be on the same page.

First, there must be connection between the intended topic to be taught (goals and objectives), and the measured outcomes (examination results). If the intended goals and objectives of the curriculum are not the same as the measured outcomes, then the curriculum will fail. Second, there must be congruence between the teachers' ideas about the curriculum and teaching strategies selected, and the expectations of the students. Absence of congruence means that the teacher and students are 'at odds' with each other. The students fail to learn and do not meet the examination criteria. In this case, the curriculum has failed.

A dysfunctional curriculum is one in which there is a mismatch or lack of congruence between its components. When a large sector of a population fails the examinations (such as in South Africa and Lesotho), the curriculum components must be examined. For example, the appropriateness of the specified goals and objectives must be critiqued in terms of students' goals and existing competencies.

The abilities of the teachers to teach the particular topic to specific students must be explored. And the examinations or tests must be checked in terms of their relevance to the specified goals and objectives, and the actual teaching strategies used by the teachers.

Student problems may include a lack of interest and/or willingness to learn a particular topic, or a lack of appropriate preparation in terms of the students' prior knowledge, attitudes and skills. The presence of students' alternative conceptions can cause learning disasters, mainly because the teacher does not know they exist within the students' conceptions. In addition, the students may not realize the significance of their prior knowledge in terms of learning (or not learning) a new topic.

An important point being made with this model of teaching and learning is that a dysfunctional curriculum is not automatically the fault of the students. The cause may lie with the selected goals and objectives, the teachers' abilities to teach, the students' preparation and willingness to learn, and the appropriateness of the tests or examinations. A failed curriculum does not necessarily mean that the students are stupid.

In the case of the failed 'Foundations of Science' course described earlier, there are several possible reasons for this dysfunction. First, the espoused curriculum was interesting to the lecturers who were all western scientists. But the students, who were all Africans from Lesotho, Swaziland, and Botswana, most likely found it neither interesting nor comprehensible. Their own indigenous knowledge (IK) was not represented in any way. African IK does exist, but the teaching team did not know about it and did not ask.

Second, we used lectures in conjunction with a student handbook with relevant readings. We also facilitated discussion groups in order to clarify the lectures, but did not find it easy to engage the students in these discussions. Somehow it was difficult to bridge the gap in the different ways of knowing despite our best intentions.

Third, the methods of science probably did not resonate with the students. The objective, reductionist and atomist line of thinking is different from the holistic way of seeing the world by Africans.

Despite our well-meaning intentions to help our students, we, the faculty, failed to take account of our students' existing knowledge, attitudes, and skills concerning describing and explaining the world in which we live. We needed to know the ideas and ways of knowing that the students brought to the lectures and discussions. And then the lecturers would have needed to know how to manage these diverse ideas and ways of thinking in the context of the course. This was (and still is) easier said than done. I venture to say that our western ignorance of our students was the main problem with this course.

Research Study Part 1: Identifying the Indigenous Knowledge of Traditional Healers in South Africa and Lesotho

While traditional healers (THs) do not usually have a role in science classrooms, and teaching children is not a part of their mission, they are, nevertheless, important and powerful people within their African communities. My hunch was that their thoughts about education might serve as a launch pad for bridging the cultural gap between western and indigenous ways of knowing. I believe that, as in the past, THs are the holders of African indigenous knowledge (IK), and curriculum decisions should include their knowledge and point(s) of view. I decided to ask THs in two different southern African countries about the ideas concerning science education in their respective countries: Lesotho and South Africa.

Science Education in Lesotho

Lesotho is a small country in the mountainous southeastern part of Africa, surrounded by South Africa. The land is largely unsuitable for farming crops, thus the people primarily raise cattle, goats and sheep. After more than 80 years of quasi colonial rule by Britain, Lesotho finally become independent in 1966. Then followed a period of foreign assistance and neo-colonial domination. However the British connection was firmly established and Lesotho embraced the Cambridge Overseas School Certificate (COSC) and examination standards.

Recently the Basotho school curriculum specified the need for teachers to include cultural and indigenous knowledge in the curriculum. The science curriculum makes some provision for IK by recommending that everyday knowledge and practices must be included in the COSC curriculum. These can include topics such as brewing of local beer, bread making, and some medical practices.

In South Africa the situation is similar. The curriculum before 1994 was western in terms of the ideas and organization of these ideas. It took no account of the IK of the local people. Recent innovations in South African science curriculum development have attempted to bridge the gap between the western and the African viewpoints by suggesting that learners should learn about 30 % of the science topics within the context of their indigenous worldview, e.g., how to make indigenous beer. But these suggestions only scratch the surface of the African IK system, which often remains obscure and even unknown to modernized (westernized) Africans involved in curriculum decisions.

THs' Ideas About Children's Learning Needs in Lesotho and the Western Cape

I wanted to learn about THs' ideas of what children should know about their world. In addition, I wanted to find out whether their ideas can be generalized across two very different countries within the African subcontinent, or specific to particular countries such as South Africa (urbanized, somewhat westernized and heterogeneous), and Lesotho (predominantly rural and homogenous). So I investigated the IK concerning health and education of both Basotho THs, who live in Lesotho, and Xhosa THs who live in the Western Cape province of South Africa.

This study aimed to find out THs' ideas concerning topics that could potentially be integrated into science teaching, and their suggested teaching approaches. The research involved two parts:

The first concerned THs' ideas about children's learning needs. My plan was to address the following questions: What are Basotho and Xhosa THs' ideas concerning the learning needs of children in health and science education? How do the ideas of the Basotho THs compare with those of the Xhosa THs in the Western Cape? And what are responses of regular classroom teachers to the TH ideas? In the second part of the study, I asked THs for their views on integrating IK in the science classroom.

I held a group interview with senior Basotho THs in Maseru, the capital of Lesotho, and with selected community THS in two rural Lesotho areas. I also sought the ideas of Xhosa THs in the Western Cape in South Africa in both individual interviews and a group session. Finally I compared the ideas from the Basotho and Xhosa groups to check the commonalities and differences between the two groups.

The interviews were either videotaped or recorded with handwritten notes. The data were analyzed qualitatively (Strauss and Corbin 1990). The THs' ideas were analyzed, and categorized into prevalent themes. Written consents were obtained for all parts of this research. Translators were employed whenever needed.

Basotho Study (see Hewson 2012)

The Basotho THs live in Lesotho, the independent African country enclosed within the geographic boundaries of South Africa. The people of Lesotho are referred to as the Basotho. A single Basotho person is called a Mosotho. Their language is Sesotho. Lesotho is mostly mountainous, with little agriculture, and is economically poor. The people are culturally homogenous and predominantly rural. Many Basotho men go to the gold mines in South Africa as miners, serving as migrant workers. For the women and children who stay behind, contact with westerners and their way of seeing the world is limited. Prior to independence in 1966, Lesotho experienced a type of English colonialism as a protectorate, but was never colonized.

Through the Department of Science and Technology in Maseru, I met and interviewed nine senior THs. They held leadership positions in the Traditional Healers Association of Lesotho and the Lesotho Initiation School Council. Some of them were chiefs, and one was also an advocate and lecturer at the National University of Lesotho. All were older men. They wore their traditional clothes including Basotho blankets, fur hats, and some carried their whisks or sticks. Translation was not needed as most spoke English well.

The head of the Traditional Healers Association then organized two meetings with a large number of community THs in the outskirts of a small town, and also in a rural area in the high mountains. During these meetings the THs first discussed their own business with accompanying singing and dancing. Then I interviewed 5 of these THs who volunteered to talk with me. They were all male except one. While the men wore their traditional Basotho blankets, the woman had a red, black and white cotton wrap, and a headdress with red and white beads. All the healers had their accoutrements (special hats, necklaces, 'whisks' and sticks). I provided small gifts to everyone (Figs. 8.1, 8.2, 8.3, and 8.4).

Western Cape Study (see Hewson et al. 2009)

The South African THs were from the Xhosa group. They speak isiXhosa and some English. They live in the peri-urban outskirts of Cape Town known as The Flats. These THs had experienced racial segregation during the *Apartheid* era of

Fig. 8.1 Leader of the
Basotho Traditional Healers
Association (Photographer:
Peter Hewson)

Fig. 8.2 Basotho Traditional Healers discuss educational ideas (Photographer: Peter Hewson)

Fig. 8.3 Tradition Healers assemble in the Basotho highlands (Photographer: Peter Hewson)

Fig. 8.4 Basotho Traditional Healer participating in discussion (Photographer: Peter Hewson)

South Africa. Most of them had little or no education. They were somewhat familiar with the western way of life—some of them had worked in white-owned businesses or households. We celebrated this coming together with a special meal, namely tripe, favored by the THs.

Methods and Data Analysis. All the interviews were analyzed using qualitative methods. I have combined the results from both studies in order to compare and contrast what the THs said in response to the questions. As an example of the ideas of the THs, I have selected one of the interviews. This Xhosa man had completed high school and was fluent in English. He was working for a domestic abuse clinic in Cape Town at the time, and was deeply concerned about sexual violence within society.

The combined data from the THs in both Basotho and Xhosa groups are summarized here in terms of seven categories:

The negative effects of colonialism and modernization. The Basotho THs who held leadership positions within Lesotho believed that the chaotic lives of the people in Lesotho are a reflection of the breakdown of their African cultural values and the difficulties they experience in assimilating western values. This chaos has led to the destruction of family life, impoverishment, and unruly, disrespectful citizens. The Xhosa THs did not mention the role of South African history, but insisted that the new South African government would need to intervene to help secure the place of IK content and THs as co-teachers in science classroom.

Xhosa Traditional Healer Suggests Children's Learning Needs
"High school students must know what a traditional healer is, and develop positive attitudes. They must learn to respect the ancestors who are always with them. They must know the difference between western healing and traditional healing. Students know about western doctors but they no longer know about traditional healers who can cure diseases.... Also students must know the 'what' and 'why' of traditional healing, e.g., traditional healing involves a calling from the ancestors. Students must know the difference between *sangomas* (THs) and witchdoctors who are called *amagqwirha*. Traditional healers cure people—they do not kill or bewitch them.

Education to prevent HIV/AIDS infection can start at puberty—from about 10 years onwards. What (young) people can do is abstain from sex.... Regular teachers have developed a western mentality that all people should have sex, whereas the African way is to disallow sexual penetration before marriage. Checking young girls (for virginity) is good. It prevents STDs and HIV. They must remain pure until married. Prevention of HIV/AIDS is also by using condoms.

Traditional healers and teachers must come together and discuss things before the traditional healers can go to the schools. They can teach about the connection between plants, animals and humans. They can teach about medicinal plants and how they work. Also they should teach about conservation (of plants). A biology teacher can teach about plants, and then the TH adds information about what the plant does in your body. I suggest the kids, about 15 of them, can sit in a circle, in a workshop arrangement. The TH can bring plants or pictures, and then demonstrate how to use them. When ideas come up, they can concretize them."

Hewson et al. (2009). Reprinted with permission from the publisher.

Main Ideas Identified from Xhosa TH Interviews
- Science is right and good and our children must learn it.
- Students must be taught to respect our African culture and traditions, especially traditional healing beliefs and practices. They should know about THs, and know how THs differ from western doctors and witchdoctors.
- Students must be taught how to care for themselves using traditional practices involving plants and animals (how to grow, and protect them, and use them for medicines), and learn about human interdependence with plants and animals.
- Students must be taught how to avoid infection, e.g., HIV/AIDS and TB, and how to take care of themselves and family members through hygiene, good nutrition, exercise, drug avoidance, and rape avoidance.
- Students must be taught about sex and its consequences such as appropriate sexual behavior (sexual abstinence), virginity checking with penalties for those who are no longer intact, and the use of condoms as protection.
- We should reject 'condomization' in the schools — it encourages early sex.
- THs and elders can go to the schools and co-teach with science teachers when appropriate, using workshops and small group discussions. They could teach with concrete examples such as stories, demonstrations of actual plants, and preparation of simple medicines.
- Teachers can visit THs to get relevant IK information before teaching. They can also visit IK centers to see medicinal plants.
- We should seek government support for THs as co-teachers in classrooms.

Adapted from Hewson et al. (2009), with permission from the publisher.

The curriculum should serve to re-introduce and reinforce the African heritage. Both groups of THs expressed this concern strongly. They felt that the schools should have this responsibility.

The curriculum should emphasize the utility of plants and animals to humans. Both groups expressed the need for students to understand the connection between themselves and their environment in terms of food and medicines for survival.

The curriculum should teach about interdependence of all living things and the need for sustainable agriculture. Both groups strongly supported the idea of mutual dependence of humans and all things. There is spiritual and material interdependence—humans are partners in the web of life as well as inanimate objects (mountains, rivers, lightning, rocks, soil, etc.).

The curriculum must emphasize healthy living and appropriate sexual practices. Both groups felt strongly that sexually transmitted infections such as HIV/AIDS are of greatest importance in Lesotho and South Africa, where the mortality rates are extremely high. In addition, both groups believe that cultural norms concerning sexual abstinence before marriage or coming of age must be reinstalled. In addition, the Basotho THs were adamant that nutritional guidelines and prohibitions were essential in terms of fostering cultural obedience as well as control of premature sexual maturation in adolescents. They favored virginity testing for unmarried girls, and teaching about condoms for boys.

Teaching methods in integrated IK-science classrooms. Both TH groups emphasized that collaboration between the THs and society's elders with classroom teachers is essential, desirable, and possible. Both groups mentioned indigenous ways of teaching that include story-telling (myths, cases), physical and cognitive games (e.g., riddles), and experiential learning (e.g., field trips to the countryside, zoos, and botanical gardens).

Research is important. Both groups were concerned that research must be done to establish health related IK as well as general IK that relates to all fields of life. They were concerned that not all TH knowledge should be taught, and that THs themselves need to discern the aspects of their knowledge that need to remain secret, and those that could usefully be shared with learners at the various levels.

Comparison of Traditional Healers' Ideas in Lesotho and the Western Cape

The interviews with THs in Lesotho and the Western Cape revealed a coherent view about what could be taught and enthusiasm to be more engaged in schooling of the children of Africa. There was considerable commonality across two different cultures despite past history, different levels of modernization, and very different ecological and socio-cultural worlds.

Despite the cultural, linguistic, geographical, social differences between these two southern African groups, the ideas the Basotho and Xhosa groups mentioned were surprisingly similar. This commonality between the groups suggests that their traditional beliefs may be generalized within African society. There is likely to be a common epistemology between the Basotho and Xhosa groups, and probably other southern African groups as well.

School Teachers' Responses to the Xhosa TH Interviews

I wanted to know how educational leaders in science teaching in the Western Cape would respond to the ideas of the THs.

Senior Teachers: In a focus group setting, a group of eight senior teachers (all actively teaching in Cape Town) responded to the suggestions of the THs. Seven were high school teachers with appointments as either head or associate head of science departments, and one was a school principal. One teacher was from a primary school, and was doing a master's degree in science education.

In the focus group the senior science teachers revealed their ignorance about THs, their role in society and their practices. They confessed their confusion between the different roles of THs and witchdoctors. They acknowledged the importance of South African children learning about the IK of all people in South Africa—the 'rainbow nation.' They said they were willing to learn about traditional African health concepts as well as other traditional sciences and healing practices found in southern Africa such as Khoi/San healing practices, different tribal healing practices, Indian Ayurvedic medicine, and old Afrikaans remedies.

The senior science teachers were enthusiastic that THs should come to the schools to share their knowledge because science teachers, like themselves, are ignorant of traditional African health concepts. They were also enthusiastic about the idea of using short videotapes augmented by question and answer sessions with THs invited to the classroom. They suggested that students could do school projects on aspects of IK.

They endorsed the idea that school children would be highly motivated to learn about IK generally. However, they debated the success of a separate IK course in the curriculum, and felt real integration would be more successful because the students would take IK more seriously.

They expressed concern about the safety of some TH practices and stipulated the need for authentication and validation of practices by an accredited organization. At the same time they acknowledged that prejudice against THs exists amongst white South Africans, and that explanations would need to be offered to parents as well as in-service training for teachers. They agreed that the South African government should be involved to support this integration.

Mid-career Teachers: We also checked responses to the THs' ideas with a group of mid-career science teachers in the Northern Cape province. They were asked to respond to the Western Cape THs' ideas gathered earlier. There were 55 in-service teachers who were engaged in a professional development course. They comprised a culturally diverse group whose home languages were IsiXhosa, Setswana, Pedi, Afrikaans, and English. The open-ended questionnaire contained the same list of items reflecting the THs' ideas.

These classroom teachers gave 100 % support to the idea that children in South African schools should be taught about traditional African healing. One teacher wrote that traditional African healing is a part of their being and religion. The teachers emphasized the need for students to study medicinal plants and how to use them in their particular regions. They saw no problems in integrating traditional African healing with western science. They suggested that comparative studies of traditional and western medicines should be done. Some acknowledged that western medicines have been thoroughly tested in laboratories, which is not the case for most traditional herbal medicines, and suggested caution. One teacher highlighted the essential difference: "Western medicine is tested through laboratories and traditional medicine is spirit gifted."

Fig. 8.5 Xhosa Traditional Healers meet to share ideas for science curriculum, Western Cape, South Africa (Photographer: Peter Hewson)

They suggested various ways of integrating IK and western science such as: demonstrating medicinal plants in the classroom; showing how each plant is processed for medicinal use; encouraging the students to do research studies; organizing field trips to the country-side and to herbal gardens that specialize in the indigenous medicinal plants (e.g. Kirstenbosch Gardens in Cape Town); and bringing THs into their classrooms.

The positive response from both the senior and mid-career science teachers was highly encouraging. Without their support the possibility of using THs to integrate IK with western science teaching could not happen. However this embryonic idea will need a huge effort by science educators at all levels, THs, parents, and the South African government (Figs. 8.5, 8.6, and 8.7).

Research Study Part 2: THs' Views on Teaching an Integrated IK- Science Curriculum

I needed more details concerning the THs' IK as well as indigenous ways of teaching. So I embarked on a further study where I aimed to elicit THs' concrete suggestions for integrating IK into science lessons (Hewson 2012). For this I again worked with the group of Xhosa THs in the Western Cape. My plan was to enable the Xhosa THs to focus on, select, and then demonstrate how to teach selected IK topics. My questions

Fig. 8.6 Xhosa Traditional Healer selects card showing Candelabra tree (Photographer: Peter Hewson)

Fig. 8.7 Xhosa Traditional Healer explains how *Vygie* plant heals a sore throat (Photographer: Peter Hewson)

were: What specific topics do Xhosa THs suggest for potential integration into the science curriculum? And how would they teach the topics if they had the opportunity?

Nine THs volunteered to participate in a nominal group session (Delbecq and VandeVen 1971). This process was facilitated by Mirranda Javu, a TH herself, and fluent in Xhosa. To start the nominal group process I used the teaching cards developed by the Western Cape Primary Science Programme (2010). These 9 × 5 in. cards have representations of the most common South African indigenous plants

(32 cards) and animals (62 cards). I used them as prompts to help the THs focus on specific and concrete biological entities. Each card has a large colored drawing of the plant or animal, its name (English, Afrikaans, and seven African languages), and its biome. All the cards were placed on tables around the room.

I introduced and explained the task to the THs. They spent the next 15 min selecting the three plants or animals they believed should be introduced into science classrooms. Their selections were shared and written on the blackboard. The duplicates were eliminated, and then each TH ranked the 3 most important plants or animals for the potential IK-science curriculum. The table shows the list of topics generated by the THs.

Topics selected for teaching topics by Xhosa THs, Western Cape

Animals		Plants	
Jackal	Lion	Sour pumpkin	Water lily
Monkey	Elephant	Wild wormwood	Baobab tree
Crocodile	Baboon	Mushroom	Aloe
Snake	Leopard	Banana	Lichen
Crab	Frog	Rooibos	Sour fig (Vygie)
Impala	Ostrich	Candelabra tree	Cotyledon
Tortoise	Millipede (Songololo)		
Tick			

Hewson (2012). Reprinted with permission from the publisher

Each TH then demonstrated how he or she would teach a selected topic in a 10-min micro-teaching session. The following table provides a summary of their micro-teaching sessions.

Micro-teaching demonstrations by Xhosa THs, Western Cape, SA

Topic	Brief outline
Crocodile	We can use crocodile skin and also its fat to heal ourselves and as a prevention against misfortune (protection from evil spirits). Crocodile skin also has economic value (belts, shoes, hand bags). Demonstrates bracelet.
Wild wormwood	Used for colds and fevers. Boil fresh (un-rinsed) leaves for 15 min, and allow to steep. Throw away the leaves. Take 2 spoons of greenish liquid every 4 h for a week (not longer). It only grows where soil is fertile.
Aloe	Aloe has juice that you can put directly into your wound, then wrap wound with a cabbage leaf. Aloe is also good for getting rid of lice and you can drink it (juice added to water) to get rid of worms. Demonstrates leaves from plant and shows picture.
Rooibos	This is helpful as an everyday tea. It's good for sleeplessness. It stimulates appetite. It is good for nappy-rash. You can boil it and use a soft towel to bathe the baby. It is important for nursing mothers to put on their breasts.
Sour fig (Vygie)	It is very helpful for sore throats and tonsils. You must chew directly on it. You can boil it with water and gargle with it 3 times daily for 3 days. It's good for babies with rash. Can use it for facial skin rash. It takes 3 days to clear pimples.

(continued)

Topic	Brief outline
Candelabra tree (*Euphorbia*)	This tree can be found in the forest. It is good for preventing witchcraft, and is medicine for twins to be used throughout their lives. When twins are unwell, both twins must wash with it. When you collect this plant you must leave a gift – a coin or a shiny thing. If you don't, the tree will cry and you will see white milk. It is not for internal use because this milk is very poisonous.
Baboon	This animal stays in the forest and eats green leaves. The *sangoma* uses baboon skin for enhancing *sangoma* dreams. It can be a *sangoma's* hat. You can wear baboon skin to become a 'big' (successful) student.
Lichen	This amazing medicine grows on rocks. It helps heal wounds that do not heal. Take lichen and burn it, then grind it, and put it on a clean wound. Take a leaf of cabbage and wrap it around the wound for 3 days. It is also good for STDs e.g., gonorrhea. Soak the lichen in boiled water then wash externally with the water, and you will be healed.
River pumpkin	This plant is good as a contraceptive. It grows by rivers but you can buy it at the market. Boil the roots for 3 min, crush them, soak them, and then cool the liquid. Drink 100 ml 3 times daily. Don't overdose – it causes abortions if already pregnant. Farmers can use it for handling cows' after-birth. You can also use it to help women get pregnant. Mix the roots with clay. When the plant starts to grow and make leaves, the water that surrounds the roots will enhance pregnancy.

Hewson (2012). Reprinted with permission from the publisher

These ideas from the THs could easily be introduced into school science teaching. As mentioned earlier, the Western Cape Primary Science Project in Cape Town produced teaching cards with pictures and basic facts about indigenous plants and animals in southern African environment. Amongst other things, each card included facts about the indigenous uses of the plants, where this was known. These teaching cards are an innovative way of showing how science teachers can start integrating IK into their classrooms.

In this chapter I have discussed a model of teaching and learning to help diagnose the ways in which a curriculum can be functional as well as dysfunctional. I briefly described the first part of a research study to find out how THs perceive the learning needs of the children in both South Africa and Lesotho. In Chap. 9, I will describe the next steps of the study in which I elicit THs' ideas about indigenous teaching strategies in South Africa and Lesotho.

We Can Teach the Children: Additional material to this book can be downloaded from http://extras.springer.com

References

Delbecq, A. L., & VandeVen, A. H. (1971). A group process model for problem identification and program planning. *Journal of Applied Behavioral Science, VII*, 466–491.

Hewson, M. G. (2012). Traditional healers' views on their indigenous knowledge and the science curriculum. *African Journal of Research in Mathematics, Science, and Technology Education, 16*(3), 317–332.

Hewson, P. W., & Hewson, M. G. (1988). An appropriate conception of teaching science: A view from studies of science learning. *Science Education, 72*(5), 597–614.

Hewson, M. G., & Hewson, P. W. (2003). The effect of students' prior knowledge and conceptual change strategies on science learning. *Journal of Research in Science Teaching, 40*, S87–S98.

Hewson, M. G., Javu, M. T., & Holtman, L. B. (2009). The indigenous knowledge of African traditional health practitioners and the South African science curriculum. *African Journal of Research in Mathematics, Science and Technology Education, 13*, 5–18.

Strauss, A., & Corbin, J. (1990). *Basics of qualitative research: Grounded theory procedures and techniques*. Newbury Park: Sage Publications.

Western Cape Primary Science Programme. (2010). *Indigenous plants & animal cards & teachers' guide* (2nd ed.). ISBN 978-0-620-40553-9. Published by W.C.P.C.P., Cape Town, South Africa.

Chapter 9
Integrating Indigenous Knowledge with Science Teaching

Abstract I describe the next steps in the research study in which I elicit traditional healers' ideas about indigenous teaching strategies. I offer a model of teaching and learning, and a more detailed description of my teaching strategy: tailored teaching. I present two examples of how indigenous knowledge could be worked into classroom science lessons. I describe a university program in the United States and a high school in Thailand that incorporate indigenous knowledge into their curriculums, and comment on the growing interest of international organizations in identifying, testing, and preserving indigenous knowledge.

Traditional science teaching approaches in schools around the world include exposition (lectures) and the ubiquitous use of worksheets. These approaches are still commonly found in science teaching despite the modern array of teaching strategies developed by science educators.

Contemporary teaching strategies in science education involve students as active participants in the learning process. These strategies are based on extensive research and on the development of theories of teaching and learning that have evolved over the last 50 years. While I have selected only a small number of studies, they illustrate the long history of this work in education: discovery learning (Bruner 1961); learner-centred learning (Rogers 1961); experiential learning (Kolb 1984); learning as conceptual change (Hewson 1981); adult learning strategies (Knowles 1989); situated learning (Lave and Wegner 1991); and collateral learning (Aikenhead and Jegede 1999). Teaching strategies include: establishing social-spiritual-cultural relational frameworks (Cajete 1994); helping students with border-crossing (Aikenhead 2001); place-based teaching (Riggs 2003); cogenerative dialogue (Shady 2014; Tobin 2006); argumentation teaching (Ogunniyi 2004); and tailored teaching (Hewson 2000).

I will not describe each of these strategies except for argumentation teaching (Ogunniyi 2004), and tailored teaching (Hewson 2000).

© Springer Science+Business Media Dordrecht 2015
M.G. Hewson, *Embracing Indigenous Knowledge in Science and Medical Teaching*,
Cultural Studies of Science Education 10, DOI 10.1007/978-94-017-9300-1_9

Argumentation Teaching Strategy

Teaching with argumentation differs fundamentally from traditional instructional approaches that still exist in old-style schools. These approaches mainly consist of exposition—the transmission of scientific facts through lectures, catechism teaching, and worksheets. The argumentation teaching approach of Ogunniyi (2004) involves students in interpersonal interactions and discussions. It encourages teachers and students to discuss ideas so that the students can externalize and verbalize their thoughts and worldviews. In this way, says Ogunniyi, the students can decide on the relative power (usefulness, or fruitfulness) of the different ideas being discussed. And the students can both recognize and appreciate the validity and importance of their own IK.

Based on his work on argumentation theory, Ogunniyi (2004) offered an explanation for how people deal with contradictory ideas. He suggested that when students encounter new knowledge, they engage in an internal argument or dialogue concerning the competing ideas. For example, when indigenous African students are faced with a western science idea (such as the concept of density where density means mass per unit of volume), they may juxtapose this new idea with their prior, everyday idea of density meaning the number of items in a given space (such as trees in a forest). The two competing ideas may be incommensurable (Toulmin 1972), or incompatible (Aikenhead and Jegede 1999), and no resolution can be obtained. In this case, the ideas may remain equipollent. Equipollence is the situation in which two competing ideas have equal cognitive power and can co-exist without logical connection. Equipollence is similar to Turkle and Papert's (1992) idea of epistemological pluralism, but at an individual level. Ogunniyi's suggestion is particularly interesting for understanding how indigenous non-westerners might incorporate science into their IK system.

Argumentation teaching mirrors scientific practice where ideas are proposed but contradicted by the evidence, and by the arguments suggested by others. However, this is not characteristic of African indigenous style of thinking. As described in Chap. 2, the African way of knowing and thinking has special attributes that incorporate values such as communality, group participation, sharing, and respecting each other. These values are reflected in the African idea of ubuntu, or human-ness. This is the belief that a universal bond of sharing connects all humanity. It can be interpreted as: A person is a person through other people. This African value competes directly with western values of individualism and competition.

Tailored Teaching Strategy

For teaching to be successful, teachers must be 'on the same page' as their students—they need to be aligned with their students. In heterogeneous classrooms, we can hope that teachers know and understand their students, including their prior knowledge (especially their IK), but this is seldom the case.

Teachers need to find out details about their students' prior knowledge as well as their expectations for the course/class. But this takes time and is difficult. Teachers who fail to ask the right questions never find out the IK of their students. This IK remains submerged but active, often inhibiting further learning (Hewson and Hewson 2003). If a teacher assumes knowledge and skills that the students do not have, there will be a misalignment in the curriculum with casualties for the students. Conversely, if the students enjoy small group discussions and stories, and the teacher prefers that students complete worksheets, there will also be a misalignment. The tailored teaching approach takes care of this misalignment. In Chap. 10, I outline tailored teaching for medical education (see also Hewson 2000). The process is essentially the same for teaching science:

Create an appropriate learning climate. The teacher first establishes and maintains a risk-free and safe teaching environment for the students. The teacher should be enthusiastic about the lesson topic, respectful of the students, open-minded about the ideas they bring to the classroom, and supportive of the students;

Be learner-centered. The teacher should include the students' ideas (knowledge, attitudes and skills) into the lesson. This requires the teacher to balance the students' interests with those of the teacher;

Use facilitation skills. Rather than relying on expository lectures the teacher should try to facilitate students' learning by helping them reconcile their existing knowledge, attitudes and skills with those of the course;

Foster self-awareness. The teacher should help students become aware of their own knowledge, skills, attitudes and feelings concerning the topic of the lesson and the way it is being taught. By inviting students to be reflective, the teacher can learn important information about the students' reactions to class activities. Additionally, the teacher needs to be aware of her own assumptions and even prejudices about non-western students in her classroom;

Tailor teaching to the students' needs. The teacher needs to adjust the lesson plan to respond to the students' educational needs and preferences by using the teaching activities described in the following Tailored Teaching model:

Tailored Teaching Model

Prepare. It is important for the teacher to orient the students to the focus of the lesson. This helps the students retrieve related knowledge from their long term memory. It also allows the teacher to create a good learning environment that is friendly, focused, unthreatening and respectful.

Ask. Before teaching, the teacher needs to find out the students' knowledge and skills on the lesson topic, using questions (such as a pop-quiz) and tasks. This helps the teacher to identify their actual learning needs, as well as their related experiences with, and interest in the topic.

(continued)

(continued)

Give feedback. The teacher should give immediate feedback (both positive and negative) to the students. Most important, the teacher must now decide how to proceed with the lesson based on the students' revealed knowledge and skills, experiences with, and interests in the topic. This requires flexibility on the part of the teacher.

Teach. The teacher offers instruction tailored to the students' needs in conjunction with the planned lesson, using a variety of instructional methods (experiments, projects, mini-lectures, demonstrations, role-plays, argumentation teaching, case-based teaching, problem-based teaching, and simulations).

Apply. The teacher helps the students apply the new ideas (knowledge, attitudes and skills) to concrete problems. For example, knowledge about the process of photosynthesis could be applied to the question of maintaining and sustaining a garden or farm crop.

Review. The teacher should check the students' acquisition of the knowledge and skills of the lesson, for example, by inviting them to summarize what they learned.

Research Study Part 3: Investigating THs' Ideas About Teaching Activities

My recent study (Hewson 2012) was to investigate THs' ideas about both their own African IK and their perceptions of how to teach IK in the science curriculum. My intention was to explore the interface between the African and western ways of thinking with potential implications for improving the school science curriculum. Parts 1 and 2 of this research are described in more detail in Chap. 8.

In summary, I interviewed traditional healers (THs) in two different countries (South Africa and Lesotho) to find out what IK they considered could and should be taught in schools, and how it could be taught. I also checked with senior and mid-career high school science teachers for their responses to the suggestions made by the THs. Their positive responses to the suggestions from THs were very encouraging. To select IK teaching topics, I invited a different group of THs to engage in a nominal group process, which gave rise to a short list of topics that could be taught in school science classes in an integrative way (see Chap. 8).

In Part 3, I give further thought to how an integrated science lesson might be planned. I offer two exemplars of integrated IK-science lessons based on the suggestions of the THs. These exemplars are intended to meet the criteria of the model of tailored teaching and learning, and the Tailored Teaching strategy (see Chap. 8; Hewson 2000).

First, I planned a lesson on crocodiles using the information provided by the Xhosa THs. Then I developed a lesson on an indigenous plant in the Western Cape called *buchu.*

Examplar: Lesson Integrating IK Concerning an Animal in a Science Classroom

Title of Lesson: Crocodiles

Level: Primary school (Grade 5 or 6).

Objectives: At the end of this unit students will be able to:

- State the main features of crocodiles.
- Describe the habitat of crocodiles.
- Explore the dangers of crocodiles, and the uses of crocodiles and their body parts.

Key Ideas:

- A crocodile is a reptile that can be dangerous, e.g., it kills animals including humans and some of its parts (bile, liver, brain) are extremely poisonous.
- Some parts of a crocodile are commercially useful, e.g., the skin is used to make shoes, belts, and bags.
- Food: the tail meat of the crocodile makes delicious dishes.
- Crocodile skin is thought by some to confer protection from danger.

Teaching Methods:

- Prepare: Students will be shown pictures and videos of crocodiles. They will visit a museum to see preserved crocodiles.
- Ask: Students will engage in classroom discussions about crocodiles including stories about interactions with, and mythology concerning crocodiles.
- Teach: Invited local people with experience of crocodiles (zoo-keepers, hunters, farmers, and/or THs) to the classroom to talk about crocodiles. Students will look at labeled diagrams of crocodiles, and record their observations (drawings, photographs with labels) of salient aspects of crocodiles and their habitat.
- Apply: Students do a project about the sustainability of crocodile farming.
- Review: Students draw a crocodile with correct labels, and report on their findings about the ecology and management of crocodiles.

Materials needed for this class:

- Pictures and videos of crocodiles.
- Access to crocodile parts (e.g., skin, teeth, skeleton), and products (wallets, purses, belts, shoes, bracelets, etc.).
- Access to a zoo, animal preserve, or museum.
- Visit by community knowledge-holders (elders, traditional healers).
- Notebooks.

Adapted from Hewson (2012). Reprinted with permission from the publisher.

Examplar: Lesson Integrating IK Concerning a Plant in a Science Classroom

Title of Lesson: Biodiversity in the Western Cape *Fynbos* biome and the national importance of *buchu*.

Level: Senior school (Grades 8, 9, or 10) in the Western Cape region

Objectives: At the end of this unit students will be able to:

- Identify and name several different local plants in the Western Cape Fynbos biome, e.g., *buchu* (*Agasthoma betulima*).
- Describe the structure of *buchu* and its natural habitat.
- Describe the uses of *buchu* as a tea for enjoyment, health and well-being, and its economic importance.
- Discuss the protection, and conservation of *buchu* in the Western Cape.

Key Ideas:

- The nature of a biome, e.g., the *Fynbos* of the Western Cape.
- The structure of *buchu*, how to identify it, and where to find it.
- The particular environmental needs of *buchu*.
- The uses of *buchu* medically (e.g., calming tonic), the beauty industry (e.g., perfumes), and food industry (e.g., teas), and its commercial value.
- The history of IK concerning *buchu* and its future uses.
- Dangers in over-harvesting *buchu*, problems of commercial exploitation, and how to conserve it.

Teaching Plan:

- Prepare: The teacher invites relevant personnel to talk about *buchu* (e.g., scientists from Medical Research Council, THs, farmers, industrialists, and/or botanists from Kirstenbosch Botanical Gardens).
- Ask: The teacher will ask the students to talk to their parents and elders about stories concerning *buchu*, and report these stories to the class.
- Teach: The teacher arranges field trips to botanical gardens or the countryside (e.g., Kirstenbosch Botanical Gardens) and an industrial *buchu* preparation site. Students draw *buchu* plants and flowers.
- Apply: Students investigate ways to enhance sustainable farming of *buchu*. They attempt to grow *buchu* in school laboratory or garden. Teacher makes *buchu* tea and allows students to taste it.
- Review: Students summarize their notes on key ideas (e.g., biomes, sustainable farming of *buchu*, and its medical/industrial uses).

Materials needed for this class:

- Relevant personnel, pictures, photos, articles.
- Notebooks.

Article about *buchu* (Xaba and Lucas 2009)

Hewson (2012). Reprinted with permission from the publisher.

Note: These two lesson plans are proposals and have not been tested in classrooms. My point is to illustrate how integrated IK-science lessons concerning crocodiles and *buchu* might take shape.

Integrating IK into a University Program

A contemporary biology teaching program at the University of Wisconsin (the Undergraduate Biology Laboratory Program) has embedded principles of cooperative teamwork, team teaching, peer partnering, and communication skills (written and verbal), within the program (Klaiwong et al. 2011). Significantly, the program also aims to utilize students' experiences, their existing knowledge, and their creativity in solving complex biological problems. The students need to learn how to be effective members of a productive, collaborative research team, and are mentored to do so by their instructors.

Integrating IK into a School

At a different level, a high school in Thailand has integrated IK into the functioning of an entire school. This serves as a remarkable example of what can be accomplished, especially when royalty intervenes and creates (and funds) a radically different educational approach.

During the years 2000–2010, I visited Thailand and was introduced to this school that integrated IK throughout the curriculum. The school is situated in a relatively poor part of Thailand, where the people are mainly farmers. During the recent past the Royal Family furthered the re-emergence of IK in society. The Queen promoted indigenous arts such as silk weaving and ancient techniques of creating national treasures in gold and silver. Bangkok houses the Thai Traditional Medicine Museum and Training Center (established in 1993) where ancient herbal medicine and massage practices are preserved and taught.

Her Royal Highness Princess Maha Chakri Sirindhorn, daughter of His Majesty the King Bhumibol, visited the Northeastern region of Thailand. She expressed her 'royal wish' to upgrade the region by establishing *Nayao* School as a Private Welfare Education School (General Education). She had the insight to realize that the students in this region were predominantly rural, poor, and attached to an agrarian way of life. For a high school to be successful it would need to incorporate the needs of the children in the area, and include their IK as much as possible. At the same time the school would use modern teaching approaches and introduce the knowledge and skills of modern science, and high-tech information processing skills.

I was invited to tour this school and found intriguing ways in which IK was seamlessly integrated into the fabric of school life. The principal told me that most of his students now pass the final certifying school examinations. Many go on to university and return back to the region to help their own people improve their standard of living. One of my Thai colleagues had volunteered as a student teacher at this school:

Nayao School

Nayao School adopted the Thai philosophy of Sufficiency Economy, which aims to develop farmers' self-sufficiency through proper management of the land, and through community development. The primary aim of *Nayao* School is to fulfill the basic needs for students (and their families) in local communities to have enough to live on, and to live for.

Indigenous knowledge and educational practices have been built into the school curriculum. The Sufficiency Economy principles have been integrated into the teaching and learning activities of every school subject. In this way, the school and community work cooperatively. To accomplish the goal of a Sufficiency Economy, the school's 24 ha area of land and natural resources have been arranged into four learning stations:

Station 1 – Co-operative Store. This is the core station and is centrally located in terms of the other sub-stations such as rice farming, vegetable and mushroom farming, chicken farming, and the processing of products (e.g., rice and sugar). Students involved with the Co-operative Store manage all the sub-stations. They are able to learn not only farming and harvesting, but also how to buy, sell and make a profit on the farm products. The Co-operative Store also offers various services such as dress-making, hair-styling, water purification, development of natural insecticides, and motor mechanics.

Station 2 – Fishery. The fishery station provides knowledge and skills for fish farming. It also provides an opportunity for students to practice basic commercial and accounting skills. In addition, the school works together with the local community by inviting indigenous people, who have particular skills and IK to teach the students.

Station 3 – Botanical garden. This was established within the adjacent jungle to promote environmental awareness and natural conservation. It is used for fieldwork in science subjects, especially Life Science and Chemistry.

Station 4 – The New Theory Agriculture. This involves the application of the philosophy of Sufficiency Economy. Students learn to follow the systematic guidelines for managing limited land and water resources.

Report by Kamonwan Kanyaprasith, Ed.D.
Science Education Center of Srinakharinwirot University

My visit to *Nayao* School revealed a successful school that seamlessly bridged the indigenous and modern world. The school had state-of-the art computers and language laboratories. There was no watering down of the curriculum, and the students were successful in the national examinations. Their learning was situated deeply in the activities of life for that part of the world. Moreover, the students were expected to experiment with, and innovate practices at every opportunity. Science and mathematics permeated everything the students did.

Nayao School is an example of how to integrate IK successfully within a western type of school. In this school, the curriculum accommodated both indigenous knowledge and skills—the intended curriculum and curriculum outcomes were contiguous. The teaching and learning approaches seemed aligned—the students spent the majority of their time outdoors. They experimented with improving the output of rice and fish farming by increasing the numbers of plants (or fish) in a given area and carefully measuring the yield. They experimented with making insecticides with the ash of wood from particular trees. They were trying to improve the methods of orchid propagation by changing some of the factors (temperature, humidity, etc.). These fairly sophisticated experiments occurred within the context of students' lives, an example of place-based teaching (See Riggs 2003). Their school science learning is situated within their known world (See Lave and Wegner 1991).

This school included the IK of Thailand in the form of knowledge and skills of indigenous farming methods, and attitudes embedded in Buddhist beliefs. The active role of meditation in the school day, and the employment of traditional practitioners (elders) in teaching about farming (etc.) maintained the social-spiritual-cultural relational framework between students in this educational system and the community in which the students live (See Cajete 1994). Most of the time students worked together in small groups that promoted collaborative learning. With teachers serving as supervisors and facilitators, the students ran the school, including cooking and serving of meals, and maintaining the school's finances, leading to student empowerment. The Thai IK was successfully integrated with modern farming concepts and practices. In terms of sustainability, the entire community benefited from this school.

Importance of IK at a National and Global Level

Approaches to, and methods of teaching non-westerners about western science are now diverse and exciting, not just for the more exotic culturally different students, but also the more subtle culturally different students within modern heterogeneous classrooms.

The World Health Organization (2003) recognizes traditional medicine, especially in the fight against HIV/AIDS. The World Health Organization has enlisted the help of THs as the first line of defense against this disease that continues to rampage through Africa. More recently, the World Bank Group (2006) recognized that IK is local knowledge unique to individual cultures and societies. They see IK as the basis for local decision-making in education and medicine as well as agriculture, natural resource management, and food preparation. Most important, they see IK problem-solving strategies as the key to dealing with the seemingly intransigent problems of poverty, starvation, and disease in Africa. The World Bank sees IK as an important contribution to global development by investigating what indigenous communities know and do, to provide a context in which to help improve local communities around the world.

This interest and involvement at the international level indicates recognition that IK has intrinsic value—a far cry from earlier mandates to stamp it out. The challenge now is for classrooms and lecture halls to learn from the World Health Organization, the World Bank Group, and other groups such as UNAID, and find ways to create two-way conceptual bridges between IK and conventional science—for the general welfare of society.

Concluding Thoughts

The teaching of science in Lesotho and South Africa continues to be a disappointing failure. The stories offered, and data reported in this book suggest that the low achievement problem may be caused by the 'disconnect' between two systems of knowing—the western and the indigenous non-western. One path toward bridging the two different ways of knowing is to locate a synergistic relationship between them. In an earlier paper I wrote:

> The successful interaction between two cultures based on mutual respect involves dialogue, the results of which may be different types of science. There should be synergism between the two cultures in which the interaction between them achieves an effect or result of which each is individually incapable. (Hewson 1988, p. 325)

So how is this synergy achieved? I will discuss the results of the research reported here in the context of ideas about curriculum, instruction, and research on IK.

Curriculum: Roberts (2007) has two visions for science teaching. He envisions an introductory science curriculum aimed to create general scientific literacy (Vision 1), as well as a curriculum for a scientific and technical workforce (Vision 2). This approach could allow IK to be assimilated into the curriculum of Vision 1. When science instruction is focused on the contexts of children's lives including their daily activities, social and geographic environments, and especially their culture, they are more likely to learn effectively (Lave and Wegner 1991). This seems to be the case in *Nayao* School in Thailand. If science teaching can include IK, science learning may become more successful. Robert's Vision I could also be applied at the secondary school levels, even at Vision 2 schools that aim to teach 'science for scientists' (Roberts 2007).

Instruction: Over the last 50 years, educational researchers have offered important new approaches to teaching science across the western-non-western cultural divide. Ogunniyi (2004) suggests that students engage in 'internal conversations' concerning competing ideas in order to find a meaningful adjustment amongst them.

He developed argumentation teaching as an approach to integrate western science with IK in South Africa. Tailored teaching strategies can help African students voice their existing indigenous ideas, and integrate them with new scientific ideas—in some way (Hewson and Hewson 2003). Sometimes ideas from different ways of knowing (e.g., African IK and western) cannot be reconciled because they are incompatible. In these situations students may hold competing ideas simultaneously—a situation Ogunniyi calls equipollence of ideas. This situation is probably more common than western researchers imagine. It certainly explains the African ways of knowing that allows for 'both/and' thinking.

Research on IK: Before IK can be included into the science curriculum, researchers need to find out more about it—both the content of IK, and how indigenous people teach it to their children. Researchers also need to find out more about the ways in which indigenous people think. Understanding the epistemological underpinnings of different IK systems remains a daunting task. As IK and indigenous teaching strategies are not well known by westerners and even by some contemporary African educators, there will need to be intense research into documenting and articulating IK in indigenous communities. The THs in my studies stated that they need to do their own research on their IK—not all their ideas can be shared in general classrooms.

The research on IK described here shows that cultural concepts between two different African groups are fundamentally similar. The traditional teaching approaches suggested by the THs in both groups are also similar. The integration of IK into conventional science classrooms is, however, not an easy proposition. The cultural diversity within a country such as South Africa makes it difficult to know what to integrate and how to do it. In a unified culture such as in Lesotho, the task is easier, but still difficult. Extensive research is needed on effective and suitable teaching approaches to incorporate IK in classrooms. Twenty years ago Cobern wrote:

> Science education is successful only to the extent that the science can exist in a niche in the cognitive and socio-cultural milieu of students. (Cobern 1993, p. 57)

This is particularly important for the education of non-western students, especially where foreign curricula have been imposed on them by colonialism, or by their own desire to become aligned with the western world in order to compete and participate in the global economy. However, educating children to become viable modern citizens takes more than simply importing foreign curricula. If the curricula are not ecologically suited to the intellectual niches of the people (especially their cultural heritage), they will fail. The students will fail in the exams and will become increasingly disadvantaged with an added self-image of inadequacy.

The narratives and research data reported in this book suggest an exciting avenue to follow concerning the introduction of IK into the science curricula of southern African schools. They offer a potentially viable mechanism for integrating IK with these curricula by calling on the THs, the holders of traditional African knowledge and wisdom, to play a more formal role. The collaborative initiative between school teachers and THs would require several steps: analyzing the IK relevant to school curricula; developing common understandings of the meanings of terms; verifying and standardizing selected IK appropriate to different school levels; developing learning objectives for the curriculum; building trust and mutual respect between teachers and THs; and mutual sharing of critical information.

Recently, Hernandez *et al.* (2013) suggested a model of culturally responsive teaching involving content integration, knowledge construction, prejudice reduction, social justice, and academic development. Their research summarizes the approaches taken by other theorists, and in particular focuses on the socio-political components of educational change concerning prejudice reduction and the need for social justice. While the THs and educators involved in my studies hinted at these two issues, much more work needs to be done. Changing the provision of education to include THs and their IK would be like changing the course of a steamer under sail. Despite the difficulty, there are exciting opportunities ahead.

Such a venture would have implications for teaching science in many countries around the world. Even so-called 'developed' nations (such as in Europe, Canada, Australia, South America, Central America, and the USA) are confronted with the challenge to teach the sciences more appropriately to the increasingly diverse students in most classrooms. The type of approach described here could have huge implications for progress in difficult global problems such as halting climate change, fostering sustainable agriculture, and improving health and well-being around the world.

We Can Teach the Children: Additional material to this book can be downloaded from http://extras.springer.com

References

Aikenhead, G. S. (2001). Students' ease in crossing cultural borders into school science. *Science Education, 85*, 180–188.

Aikenhead, G. S., & Jegede, O. J. (1999). Cross-cultural science education: A cognitive explanation of cultural phenomena. *Journal of Research in Science Teaching, 36*, 269–287.

Bruner, J. S. (1961). The act of discovery. *Harvard Educational Review, 31*(1), 21–32.

Cajete, G. A. (1994). *Look to the mountains: An ecology of indigenous education*. Durango: Kikavi Press.

Cobern, W. (1993). Construction of knowledge and group learning. In K. Tobin (Ed.), *The practice of constructivism in science education*. Hillsdale: Lawrence Erlbaum.

Hernandez, C. M., Morales, A. R., & Shroyer, M. G. (2013). The development of a model of culturally responsive science and mathematics teaching. *Cultural Studies of Science Education, 8*, 803–820.

Hewson, P. W. (1981). A conceptual change approach to learning science. *European Journal of Science Education, 3*, 731–743.

Hewson, M. G. (1988). The ecological context of knowledge: Implications for learning science in developing countries. *Journal of Curriculum Studies, 20*, 317–326.

Hewson, M. G. (2000). A theory-based faculty development program for clinician-educators. *Academic Medicine, 75*(5), 498–501.

Hewson, M. G. (2012). Traditional healers' views on their indigenous knowledge and the science curriculum. *African Journal of Research in Mathematics, Science & Technology Education, 6*(3), 317–322.

Hewson, M. G., & Hewson, P. W. (2003). The effect of instruction using students' prior knowledge and conceptual change strategies on science learning. *Journal of Research in Science Teaching, 40*, S87–S98.

Klaiwong, K., Hewson, M. G., & Praphairaksit, N. (2011). Identifying pedagogical content knowledge in an undergraduate inquiry biology laboratory course. *International Journal of Learning, 17*, 17–30.

Knowles, M. S. (1989). *The making of an adult educator.* San Francisco: Jossey-Bass.

Kolb, D. (1984). *Experiential learning: Experience as the source of learning and development.* Prentice Hall: Englewood Cliffs.

Lave, J., & Wegner, E. (1991). *Situated learning: Legitimate peripheral participation.* Cambridge, UK: Cambridge University Press.

Ogunniyi, M. B. (2004). The challenge of preparing and equipping science teachers in higher education to integrate scientific and indigenous knowledge systems for their learners. *South African Journal of Higher Education, 18*(3), 289–304.

Riggs, E. M. (2003). Field-based education and indigenous knowledge: Essential components for geoscience teaching for Native American communities. *Science Education, 89*, 296–313.

Roberts, D. (2007). Scientific literacy/science literacy. In S. K. Abell & N. G. Lederman (Eds.), *Handbook of research in science education.* Mahwah: Lawrence Erlbaum.

Rogers, C. R. (1961). *On becoming a person.* Boston: Houghton Mifflin & Co.

Shady, A. (2014). Negotiating cultural differences in urban science education: An overview of teacher's first-hand experience reflection of cogen journey. *Cultural Studies of Science Education, 9*(1), 35–51.

Tobin, K. (2006). Aligning the cultures of teaching and learning science in urban high schools. *Cultural Studies of Science Education, 1*, 210–252.

Toulmin, S. (1972). *Human understanding, Vol. 1: The collective use and evolution of concepts.* Princeton: Princeton University Press.

Turkle, S., & Papert, S. (1992). Epistemological pluralism and revaluation of the concrete. *Journal of Mathematical Behavior, 11*(1).

World Bank Group, (2006). Traditional medicine programmes in Madagascar. Retrieved January 1, 2014. http://www.worldbank.org/afr/ik/iknt91.pdf

World Health Organization. (2003). *Report on WHO traditional medicine strategy 2002–2005.* Geneva: World Health Organization.

Xaba, P., & Lucas, N. (2009). Buchu. *Veld & Flora,* September, 160–161.

Chapter 10
Incorporating Indigenous Knowledge into Clinical Teaching

Abstract In trans-cultural clinical teaching, differences in language, expectations, or cultural (indigenous) knowledge, beliefs and skills can cause problems. Medical students with such problems are challenging to their medical teachers. Clinical teaching offers many opportunities in which indigenous knowledge can be revealed and managed appropriately to facilitate learning. One way is through the Tailored Teaching approach.

In South Africa the overwhelming majority population is African, yet western medicine is the officially accepted approach to, and standard of care. Even though the vast majority of Africans still visit traditional healers for everyday problems, the traditional African approach to health and healing was never officially sanctioned by the previous western *Apartheid* government, and is still not quite acceptable in the 'new South Africa.' The World Health Organization (WHO 2002) has, however, recognized the potential of traditional medicine as an adjunct of western medicine, especially in the fight against AIDS. In any case, the need for trans-cultural medical understanding is great.

Medical Care

Trans-cultural medicine must deal with different approaches to medical care among different cultural groups and subgroups. It involves not only the physical aspects of health care, but also social, psychological, and religious influences. It concerns divergent views about health maintenance, preventive medicine, illness, therapeutic treatments, coping methods, and beliefs and practices regarding birth and death.

In addition, trans-cultural medicine concerns cultural differences between western medical teachers and their students from non-western countries. Communication between English-speaking medical educators (the doctors) and medical students representing numerous African languages is often problematic.

The American Academy on Communication in Healthcare (AACH) is an organization dedicated to reinstituting relationship within the healing enterprises of

© Springer Science+Business Media Dordrecht 2015
M.G. Hewson, *Embracing Indigenous Knowledge in Science and Medical Teaching*,
Cultural Studies of Science Education 10, DOI 10.1007/978-94-017-9300-1_10

contemporary medicine. For more than 30 years AACH has promoted research and education in relationship-centered healthcare communication. Modern medicine, while increasingly scientifically driven and technically effective in curing humans, seems to have lost its heart – the caring component of healing.

This organization focuses primarily on communication skills, both verbal and non-verbal. It encourages patient-centered care, as opposed to doctor-centered, top-down caring approaches. Members focus on both the science of linguistics and socio-linguistics, as well as the handling of emotions. They have developed measures of patient satisfaction, both within western culture and between cultures, and have focused on many issues in intercultural medical care giving, including language, culture, history, and religion.

In many ways the work of AACH brings western care giving more into line with the African approach. A fundamental premise of African IK is the worldview of *Ubuntu*. This is the view that all beings are related to each other and to the cosmos. Humans are profoundly related to, and dependent on each other and all living creatures. *Ubuntu* can be summarized by the maxim: 'a human being is human because of other human beings,' or 'I am because we are, and we are because I am' (Chuwa 2014, p. vii). The *Ubuntu* principle suggests how people ought to behave (axiology, morals). *Ubuntu* preserves harmony between people and promotes reciprocity between them. It also promotes solidarity between people that unites individuals and the community they live in. This principle underwrites the way in which THs engage with their clients. For example, trust between patients and caregivers is of utmost importance. THs do not proceed with a patient's care if they detect distrust or insincere motives. In the western context, medical distrust has been associated with reports of lower quality of communication between patient and caregivers, contributing to healthcare disparities in situations involving racial or ethical differences (Schwei et al. 2014). See Chap. 5, p. 71 for a discussion on an integrative approach to patient care

Medical Teaching

Just as there is divergence between doctors and their patients from different cultures, there is also tension between western medical teachers and their students from different cultures. Misunderstanding between the medical doctors and their students reflects differences in their respective values and beliefs. These misunderstandings can cause anxiety, distress and even anger in the student. There can also be disappointment or frustration for the medical teachers, especially clinical teachers. The following narrative comes from an amalgam of two different medical students I met while teaching in South Africa. Using creative non-fiction, I have assembled their stories into one.

Thabisa – Medical Student, 1984

With strong encouragement from her parents, Thabisa applied for medical school in South Africa and was accepted. She had done well in her high school, and her parents had great hopes for her. The family came from a rural part of the country. Her cultural group was Basotho, and her home language was Sesotho. With the new South African system of instituting racial and economic equality her father rose in the ranks of a mining corporation. Both parents wanted the best for their daughter and medicine seemed an appropriate route for her. She started medical school with trepidation but also with the determination that got her through science and mathematics during high school.

Within a short time Thabisa started falling behind her classmates. Her intellectual skills were not the problem – as long she was calculating and memorizing, she was fine. Her difficulties started in the obstetrics and gynecology section of the syllabus. Thabisa felt uncomfortable with the lectures detailing male and female sexual anatomy. Was this information that she needed to know? Wasn't it taboo to be talking about such personal things? In African society children never see their parents naked, and even in marriage, it was common practice that wives did not actually see their husband's genitalia.

When medical classes discussed sexual function and dysfunction, Thabisa was silent. In case-based problem solving about infertility she contributed nothing. Later on in her training she was required to talk to women patients about menstruation, puberty rites, dysmenorrhea, sexual intercourse and giving birth. During her first year she had to accompany a pregnant mother as she progressed through her pregnancy. Thabisa's acute embarrassment translated into reluctance.

Thabisa also seemed reluctant to take a sexual and psychosocial history. And she would not pursue questions about domestic violence. Broaching the topic of violence in the home was anathema for Thabisa. While violence against women is commonplace in many African societies, talking about it seemed wrong to her. When she was asked to make a presentation on a case involving sexual matters, Thabisa became disorganized and overwhelmed. She was unable to generate hypotheses, and often jumped to inappropriate diagnoses. She was greatly confused over the demands of her medical training and her own social mores. Her confusion turned her to stone.

Her teachers rated her as having poor communication skills, a lack of experience to perform basic skills, as well as making errors. They observed her reluctance to pursue key clinical data. Her teachers also rated Thabisa as noncompliant with assignments and tasks. When Thabisa started arriving late for hospital rounds, she was chastised and threatened with failure.

(continued)

(continued)

Being brought up in a traditional home in a rural region, Thabisa had learned that only old women (post-menopausal) were allowed to deliver babies and become involved with matters concerning sex because of the problems of *fisa*.* According to this Basotho belief, women in childbirth or menstruating are considered to be *fisa*, the condition of being hot. When people interact with women during birthing or menstruating, they become susceptible to spiritual illness and even death. Post-menopausal women are immune from this threat. Thus, as a young woman, Thabisa believed that she could not engage with matters concerning sex, birth, or death, at the risk of serious consequences for herself.

Thabisa was aware of her problem but she felt she could not explain it to her medical school faculty, who were mostly not Africans.

(Constructed case based on author's interactions with two students)

---

*See Chap. 4 for further information about this cultural belief (Hewson and Hamlyn 1985)

The story of Thabisa is realistic in trans-cultural medical teaching situations. Her first problem was her halting English – her home language was Sesotho and she found herself unable to keep up with the dialogue in lectures and group discussions. As her rural grandmother kept reminding her, she had not received a spiritual 'calling' to be a doctor, but was obeying the wishes of her urban parents. Her cultural identity was somewhat confused between the wishes of her parents and those of her grandmother. Thabisa remained voiceless in this domestic debate, yet she entered medical school with intentions to do well and get a medical degree.

As a dutiful young Basotho woman, Thabisa, was bound to follow the advice of her parents and teachers. She saw them as appropriate authorities in her life and followed their injunctions. Her preferred learning style was to memorize by rote. This approach served her well for learning the large amount of information appropriate to the first year of medical school. Her difficulties started in the practical parts of the medical curriculum. The diagnostic reasoning process was unfamiliar to her. She had little understanding of evidence-based diagnosis, and was uncomfortable with medical uncertainty. Case-based teaching mystified her, and all she wanted was to settle into the library with medical textbooks and a clear learning agenda.

Thabisa subscribed to her cultural beliefs and values. She was confused by the competitive nature of medical school. She could not understand why this was so prevalent. It made her uncomfortable. Being one of the very few black women in the class caused her to feel vulnerable and unsure of herself. She tried to be inconspicuous and became even more quiet and withdrawn than normal.

Her teachers interpreted her style as being due to her ignorance and unwillingness to participate. But when they talked with her one-to-one, they found her to be impressively knowledgeable. They experienced Thabisa as an enigma, and were unsure of whether or not to allow her to pass the semester. Much time was spent discussing her case. The medical teachers were concerned that they should not fall into the trap of cultural stereotyping.

Challenging Students

Students can become challenges for their clinical teachers for numerous reasons. One of these is that their learning may be compromised due to competing ideas (such as indigenous beliefs). Yellin (2014) discussed learning problems in medical education that are usually described as learning disabilities (discrepancies between cognitive potential and academic performance). He explains that the difficulties may not be due entirely to aptitude problems with the students, and that the medical educational system must bear some of the responsibility for the students' failure to learn. All students selected for western medical schools have already demonstrated superior ability to learn, and passed their medical examinations (e.g., MCAT). The goal of medical educators is to find out why students are struggling and provide them with tools to succeed in medical school. Most of these problems arise from inappropriate learning styles, learning strategies and study skills. Yellin (2014) suggests that neuroscience should be used to identify the nature of students' learning problems in medical schools. However, I prefer to focus on the student whose knowledge, attitudes and skills, and learning style differ from those of western students for socio-cultural, linguistic, spiritual, historical, and philosophical reasons.

Because of the intensely competitive selection process for entrance to medical school, intelligence is seldom the problem. Students from non-western intellectual environments can also be challenging to their teachers due to their different behaviors, habits, beliefs, and values. Cultural factors play into these difficulties.

In the story of Thabisa, her culture promoted behaviors that both helped and hindered her. Her willingness to simply commit to memory the massive amount of medical information provided in the first year medical curriculum helped her. She became knowledgeable and easily able to pass tests requiring factual information. Her inability to do diagnostic reasoning and the verbal problem-solving skills required in the case-based curriculum as well as during teaching in the clinic clearly hindered her. Her cultural background encouraged her to be polite and considerate to patients, and also deferential to authority. Also, her culture discouraged her from engaging in discussions involving sex, and matters of birth and death. She would have responded well to medical conversations that were holistic and spiritually informed, but these did not take place in the western medical curriculum.

Trans-cultural Medical Teaching

How is a medical teacher to communicate effectively with students across a cultural divide and avoid the traps of misunderstanding? It is difficult to know all the knowledge, beliefs and values of every cultural group. Moreover, there is cultural diversity within groups to such an extent that it is probably inappropriate to make cultural assumptions about anyone. A useful approach would be to assume that every student brings a particular world view and a specific set of cultural, social, psychological and economic perspectives to the medical education encounter.

Ausubel (1968, p. vi) recommended that before teaching, teachers should first find out what their students know, believe, and what skills they have learned. This recommendation is intuitively obvious to the casual observer, but it continues to be elusive and difficult to attain. It is, however, particularly pertinent when teaching medicine trans-culturally. According to Ausubel's maxim, the task of the medical educator is to explore each student's perspective(s) before deciding on the best way to teach a particular topic.

It is impractical to expect medical teachers to know everything about the cultural bases of all their students. What is needed is an approach that handles trans-cultural complexities in a general way. Several models provide a generic approach to take care of such difficulties, such as the developmental model of ethno-sensitivity (Borkan and Neher 1991), the negotiation model for the doctor-patient relationship (Botelho 1992), and the patient-based approach to trans-cultural primary care (Carrillo et al. 1999). I have developed a Tailored Teaching approach for trans-cultural medical education encourages medical students to reveal their perspectives, opinions, beliefs and attitudes (Hewson 2000).

Medical teaching takes several different forms. Lectures are a classic standby in medical schools, especially in the form of 'grand rounds.' Laboratory work (such as anatomy dissections, pathology and patho-physiology) is a standard teaching approach. Hospital rounds involve a clinical teacher and a group of students that visit inpatients, and gather around the inpatient's bedside for instruction. Teaching in the outpatient clinic involves a medical student working one-to-one with a medical doctor, either in a hospital or community clinic. Medical teachers (lecturers, laboratory directors, medical rounds teachers, and clinical teachers) are obliged to teach to a specified curriculum (with goals and objectives), using a variety of instructional methods. But this is not always easy for the teachers of medical rounds and clinic teaching due to the unpredictable nature of the illnesses manifested by the available patients. Despite these vicissitudes, medical teachers should teach in a way that is tailored to the students' needs. Each of the following approaches can incorporate issues concerning trans-cultural medical teaching.

Western medical educators can improve their teaching effectiveness with non-western medical students by acknowledging their non-western students' world-views and perspectives. This requires tailoring their approaches to the contexts of their students' particular environments (physical, cultural, philosophical, spiritual or socio-economic), stage of life, and life-styles. For a student to understand and accept a teacher's advice, it is necessary that he or she is able to make sense of the advice, and understand it in the context of his or her own cultural beliefs. Students need to be able to appreciate how their own knowledge and beliefs are relevant to the clinical situation, and how they could improve clinically by integrating the medical teacher's advice (Carrillo et al. 1999).

Tailored Teaching Approach

The Tailored Teaching approach has five steps that have been shown to be effective in medical education generally (Hewson 1993; Hewson and Copeland 1999). These steps offer a useful way of communicating with students from diverse cultural backgrounds. The teacher must find out the current needs of the student in order to enhance learning in clinical situations, whether in academic case discussions, or in real clinical engagements. Thus, following Ausubel's maxim, teachers need to find out what their students already know and believe before they teach. In this respect, teaching science and medicine are similar enterprises.

As a summary of the more detailed description provided in Chap. 9, I provide a brief account of the main points. Medical teachers need to: create an appropriate learning climate for their medical students, residents and fellows; they should foster the students' self-awareness, as well as their own; before teaching anything, they should elicit students' knowledge, attitudes and skills concerning the topic; they should do this in a way that is learner-centered, i.e., work from the students' established needs; and to do this they should use facilitation skills rather than teacher-centered didactics.

For this approach to be applied in medical teaching it is necessary to give some thought to the diverse range of teaching opportunities in medicine. Not all teaching opportunities in medicine are conducive to Tailored Teaching approach. The following table lists some of these teaching opportunities:

Teaching Strategies for Clinical Teachers

Strategy	Description
Exposition (lecture or mini-lecture)	Formal, logical presentation of new ideas (See Copeland et al. 2000)
Demonstrations	Effective for teaching clinical procedures & communication skills
Questions, e.g., open-ended, closed ended, exploratory, Socratic*	Direct or indirect questions to elicit students' ideas on a topic, especially IK. Encourages critical thinking as well as remembering relevant information. Reveals students' IK
Literature search and independent study	In response to a student's identified needs, ask him to search for information in journals, medical textbooks, and digital online resources, as well as mini-experiments
Case- and problem-based teaching*	Use of clinical cases and problems to stimulate active learning and small group interaction. Reveals students' IK
Role-plays, computer simulations, high-tech mannequins*	Real or standardized patient (actor) enacts a clinical case or problem, for small group discussion. Also computer-based simulations, and high-tech mannequins for surgical & anesthesia students. Reveals students' IK
Reflective strategies and personal awareness*	Videotapes of student's clinical performance fosters self-awareness of knowledge, attitudes and skills about a medical incident. Reveals students' IK (Hewson 1991; Hatem 2014)
Discussions, debates, and argumentation*	Encourages students to discuss, justify, and defend their ideas. Important for managing students' IK (Hewson and Ogunniyi 2011)
Hospital rounds and outpatient clinics (clinical apprenticeships)	Expert clinicians model, coach, articulate, and reflect on what is happening, providing clinical support, and exploring clinical possibilities (Collins et al. 1987). Offers opportunities for dealing with IK

Clinical teaching occurs throughout the hospital. My personal experience is that medical students, residents, and fellows prefer clinical teaching because it is 'hands-on' and close to the practical side of medicine. Clinical teaching occurs on the hospital ward (called 'rounds'), in the operating room, and in the outpatient clinic. I will focus on outpatient clinic teaching. In addition, I will apply the strategy of Tailored Teaching to make the case that clinic teaching need not be as unstructured as it often seems.

Tailored Teaching Approach in the Clinic

The Tailored Teaching has six activities: prepare, ask, give feedback, teach, and review. These are relevant to teaching in general, and especially to clinical teaching. Teachers may find it useful to consider these activities:

Prepare. Students from non-western cultures need to feel welcomed by the clinical teacher. Creating a warm and friendly teaching environment is a good way to start. Ensure that the students know the focus of each teaching event. Explanations of teaching methods help. Verbaland nonverbal language (vocabulary, pace, and clarity) should be adjusted to the students' needs – their English will improve. Their willingness to participate verbally and actively will also improve when they realize the lack of risk from doing so. Positive responses to their ideas from the clinical teacher and other group members will also help. Positive and supportive negative feedback will happen later. Check on the students' social context (living environment, support system, financial status, stressors, etc.). This time of transition is difficult for students from other cultures.

Ask. Use open-ended questions to find out what the students know and believe about particular clinical conditions. This would not mean a formal case presentation, but rather open questions about their views. For example, with Thabisa, the clinical teacher could enquire about cultural practices concerning birthing in her home environment in Lesotho. Ask about the student's concerns, if any. Ask about her 'explanatory model of illness.' Her answers will be informative for the medical teacher and other students in the group.

Give Feedback. Decide how to respond to the student's cognitive and behavioral needs in the context of the clinical presentation. Thabisa's reluctance to talk about sexuality, birthing, and death was cultural rather than a lack

(continued)

(continued)

of aptitude. The clinical teacher could note the difference and show interest in Thabisa's perspective. Give her feedback (positive and constructive negative) about her contribution.

Teach. The clinical teacher should describe his or her perspective on each clinical situation and offer reasons. Whenever possible use supporting literature. Make suggestions for how the students can improve their knowledge and skills. Invite students to make their own suggestions for their improvement. Thabisa's difference in knowledge, beliefs, and approach to a topic involving sexuality presents a serious challenge for a clinical teacher. Teaching approaches relevant for this student could involve empathetic listening. Both the teacher and medical students should discuss their ideas in a non-threatening, supportive environment. Then they can decide on the relative power of the different ideas being discussed. If the medical student cannot accept the western viewpoint, it is sufficient that she appreciates that there are alternative ways to view a situation. Another approach is to provide options to the student so that she could also honor her indigenous beliefs.

Apply. To cement the student's learning of new ideas the clinical teacher can encourage a students to apply the new conception to concrete facts or events. The clinical teacher could discuss 'what if' situations, for example: What if Thabisa was practicing medicine and a patient went into labor? What would she do to handle the situation? How could the student handle the medical education requirement that she do a learning experience involving maternity and birthing? Could adjustments be made to the curriculum? How could this student negotiate such an adjustment?

Review. Before concluding the teaching session the clinical teacher should ask the students what they have learned. This provides a review of the discussion, and allows the clinical teacher to find out the knowledge, attitudes and skills that were absorbed by the students. Ask each student to summarize what he or she has learned during the session. Alternatively, the clinical teacher can do the review for the students.

I exemplify this Tailored Teaching approach with a constructed narrative of an African medical student working with his clinical teacher (preceptor) in the clinic.

Exemplar – Using Tailored Teaching in the Outpatient Setting

Clinical teacher: When you have interviewed the patient, I would like you to give me the short presentation (prepare).

(Later, after the patient interview).

Student: I interviewed a 55-year-old woman who presents with nasal congestion and a headache that started 20 days ago. She says she's feeling worse. She has no fever but her ears are a little blocked. She is taking decongestants and aspirin. This is her fourth attack this year.

Clinical teacher: So what do you think is going on? (ask).

Student: It sounds like she is getting sinusitis. But I wonder if it may be an allergy. I think it is also bad luck to get sick so often.

Clinical teacher: How can you tell if it is sinusitis or allergy? (ask).

Student: We only have symptoms without any greenish mucus, which is evidence of infection.

Clinical teacher: So what do you want to do? (ask).

Student: I'd suggest antihistamines, and if there is no improvement we could try antibiotics.

Clinical Teacher: Anything else? (ask)

Student: I think this lady must also talk to her ancestral spirits to find out why she is sick so much.

Clinical teacher: Umm, let's talk about that idea in a minute. But first, the antihistamines sound like a reasonable plan. Your plan includes good management strategies for when you don't know exactly what's going on with the patient. Also, your presentation was clear and complete (positive feedback). When you mentioned the possibility of allergy in your differential diagnosis, you missed a couple of possibilities (negative feedback). You should also be concerned about rhinitis and especially the common cold. Let me explain…. (teach – mini-lecture). Now, what is this about spirits? (ask).

Student: My people at home believe a patient's bad luck may be because of bad relationships with other people, or with the recently departed spirits. If she has offended any of her ancestral spirits she must make peace with them. She could pray to them to ask for good health.

Clinical teacher: Your concern about the spirits is interesting and I can understand that in your African worldview of talking to the spirits could relieve stress and tension and promote good health. What is your justification for this idea?

Student: We have often seen how our family members get physical relief from praying to their spirits, especially in times of illness.

Clinical teacher: That's an interesting view, and I would like to hear more later on. In the USA we stick to evidence-based medicine – information based on concrete evidence from clinical trials, (teach – exposition).

Clinical teacher: OK, let's review this case.

Student: We will offer antihistamines and also check the patient for rhinitis (review).

This medical student was comfortable revealing his IK concerning the reasons a person may become ill. The clinical teacher did not dismiss it but invited him to talk further about it.

A clinical teacher will not want to enter a debate over every discrepant idea between indigenous and western medical knowledge. By allowing a topic to become part of the discussion between the clinical teacher and student, the cultural issue becomes open for discussion without criticism or disparagement. Allowing a medical student to verbalize his cultural beliefs can ease potential tensions involved in the cultural divide between western and indigenous knowledge. While it may seem to take additional time to teach in this way, the gain in student learning is maximized, particularly across the cultural divide.

We Can Teach the Children: Additional material to this book can be downloaded from http://extras.springer.com

References

Ausubel, D. P. (1968). *Education psychology: A cognitive view.* New York: Holt, Rinehart and Winston.

Borkan, J., & Neher, J. O. (1991). A developmental model of ethnosensitivity in family practice training. *Family Medicine, 23*(3), 213–217.

Botelho, R. J. (1992). A negotiation model for the doctor-patient relationship. *Family Practice, 9,* 210–218.

Carrillo, J. E., Green, A., & Betancourt, J. R. (1999). Cross-cultural primary care: A patient-based approach. *Annals of Internal Medicine, 130*(10), 829–834.

Chuwa, L. (2014). *African indigenous ethics in global bioethics: Interpreting Ubuntu.* Dordrecht: Springer.

Collins, A., Brown, J. S., & Newman, S. E. (1987, January). *Cognitive apprenticeship: Teaching the craft of reading, writing and mathematics* (Technical Report No. 403). BBN Laboratories, Cambridge, MA. Centre for the Study of Reading, University of Illinois.

Copeland, H. L., Longworth, D. L., Hewson, M. G., & Stoller, J. K. (2000). Attributes of the effective medical lecture: Results of a prospective validation study. *Journal of General Internal Medicine, 15*(6), 366–371.

Hatem, D. (2014). Chapter 14: The reflection competency: Using narrative in remediation. In A. Kalet & C. L. Chou (Eds.), *Remediation in medical education: A midcourse correction.* New York: Springer.

Hewson, M. G. (1991). Reflection in clinical teaching: An analysis of reflection-on-action and its implications for staffing residents. *Medical Teacher, 3*(2), 179–183.

Hewson, M. G. (1993). Patient education for conceptual change. *Journal of General Internal Medicine, 8,* 393–398.

Hewson, M. G. (2000). A theory-based faculty development program for clinician-educators. *Academic Medicine, 75*(5), 90–93.

Hewson, M. G., & Copeland, H. L. (1999). Outcomes assessment of a faculty development program in medicine and pediatrics. *Academic Medicine, 74*(10), S68–S71.

Hewson, M. G., & Hamlyn, D. (1985). Cultural metaphors: Some implications for science education. *Anthropology & Education Quarterly, 16,* 31–46.

Hewson, M. G., & Ogunniyi, M. B. (2011). Argumentation-teaching as a method to introduce indigenous knowledge into science classrooms: Opportunities and challenges. *Cultural Studies of Science Education, 6*(3), 679–692.

Schwei, R., Kadunc, K., Nguyen, A., & Jacobs, E. A. (2014). Previous negative health care experiences and institutional trust in three racial/ethnic groups in the US. Proceedings of the 2013 international conference on communications in healthcare. *Medical Encounter, 27*(2), 21–22.

World Health Organization (WHO). (2002, May 16). *WHO launches the first global strategy on traditional and alternative medicine.* Retrieved January 1, 2014. http://www.who.int/media-centre/news/releases/release38/en/#

Yellin, P. (2014). Earning differences and medical education. In A. Kalet & C. L. Chou (Eds.), *Remediation in medical education: A midcourse correction.* New York: Springer.

Part V
Finale

Chapter 11
Epilogue

Abstract Seeking an in-depth understanding of an indigenous world so very different from my own, I was initiated into African healing in Zimbabwe. I reflect on the meanings of being a teacher and being a medical doctor.

I had an opportunity to visit Zimbabwe to learn about African IK and the indigenous training process. I jumped at the chance. My mentor was Mandaza Augustine Kandemwa, a traditional healer (TH) who lives in Bulawayo. Mandaza was trained as a teacher, but was conscripted to serve as a police school administrator. During this time he was afflicted with severe 'water spirit' disease. While ill he dreamed of becoming an Ndebele traditional healer (TH). His job transferred him to the Ndebele region where he was trained and initiated into the community of THs. Mandaza then became a full time TH and peace-maker, working in Bulawayo.

I previously met Mandaza at an American conference, and he invited me plus three others to visit him and stay with his family. We traveled to Bulawayo in 2006. Coming from different perspectives, each of us was interested in African healing, and Mandaza promised to teach us. He opened his door and heart to us, and we joined in the daily routines of his large family—his wife, many sons and their several wives, and numerous small children. Each day his clients came to the house—some for healing and others for training as THs.

We became full members of Mandaza's household, seeking to learn the knowledge and skills of indigenous healing through full participation in the training process.

Zimbabwe, 2006

The small plane skittered to a stop on the empty runway at Bulawayo airport, the second largest city in Zimbabwe. As I entered the corrugated iron terminal building I noted the officials in khaki uniforms with dark glasses, and guns.

What was I doing here? What if a coup against Zimbabwe's strongman President Robert Mugabe had started? The possibility of social mayhem made me sweat. I wasn't sure who would fetch me, where I would stay, and I had no Zimbabwean currency. And here I was, heading into an experience of African

(continued)

© Springer Science+Business Media Dordrecht 2015
M.G. Hewson, *Embracing Indigenous Knowledge in Science and Medical Teaching*,
Cultural Studies of Science Education 10, DOI 10.1007/978-94-017-9300-1_11

(continued)

living and spirituality. I wanted to learn all I could about African thinking and needed to submerge myself in it. There are things you cannot learn from books—you have to be there, get involved—body, mind, and spirit.

The airport was beginning to close down for the evening. Would I have to sleep here on these awful metal chairs? I was relieved when one of Mandaza's sons came to fetch me. Soon I was at the family house in the Bulawayo suburbs. Three friends from the US had already settled in: a medical doctor, a shaman and a writer. The four of us were on spiritual quests, each in our own way.

"Come in, come in! You're very welcome."

A group of four small children danced and sang *Shosholoza*, a South African (Zulu) freedom song. Could this visit be a freedom from western thought patterns for me? I wanted to be able to slip into the African way of thinking and being, and transition from my usual western mindset. In this way I could begin to understand the difficulties that African children experience when learning science in their western oriented schools. I suspected these difficulties were caused by something deeper than lack of resources, perhaps by a fundamentally different way of seeing the world. And how could I know this if I didn't put myself in an equivalent position, with me trying to learn about their world?

Mandaza showed me his garden with plantings of local indigenous plants used for medicinal purposes. "I'm growing a plant that we healers believe will help patients with AIDS. It's a project with the Zimbabwean Medical Council." I felt encouraged to find synergy between the western and African worldviews. But I knew full well that western biochemists were mostly interested in the chemical components of African indigenous plants. They had no interest in the spiritual side of healing—it was not within their scope. Rather they were eager to exploit these powerful healing agents if they could find them. I had come to believe both the scientific and spiritual components were needed for understanding the power of African medicine.

"Come inside, I want to show you my sacred hut." Mandaza and I stooped to pass through a half door with a wire loop for a latch. A musty smell permeated the dark shadows. All his special things were here. Countless bead necklaces hung on wires from the roof, as well as several snake skins. There were carved sticks propped against the wall, as well as some sharp knives or perhaps they were spears. Cured *jackal* skins covered the two small chairs. A large throne-like chair was draped with a leopard skin and beads—Mandaza's chair.

I experienced a sense of energy in this place. Was it like a chapel? A shrine? Perhaps it was more like a medical clinic. Perhaps none of these analogies worked, and trying to interpret the place in terms of my western frame of reference was not appropriate.

(continued)

(continued)

I also felt apprehensive. Was there some unspecified danger with so many dead animals and killing instruments lying around the hut? I was unsure about fully immersing myself in this vastly different spiritual life. Could I maintain my equilibrium? Would my sanity be at risk? I was struggling with the appropriateness of getting tangled and enmeshed in the African way of seeing the world. An anthropologist friend of mine had warned me against doing this sort of immersing. And here I was, about to take a swim in a dark pool of unknown dimensions, and of unknown outcomes. I needed to let go of my objectivity, and surrender to the mystery. Willingness was one thing, but doing it was quite another. I was concentrating on being open-minded and curious about whatever was presented to me.

The floor was covered with soft clean river sand. A lion skin, complete with its head, lay on the sandy floor. Several large river stones shone white in the darkness. Mandaza knelt down and lifted a plate off the floor. Under it was a small well, or perhaps a reservoir.

"This is the sacred water of the Water Spirits." He scooped out some with a soup ladle to show me. "If you like, you can spend a night here. You can rest on the lion skin and lean your head against the lion's head. You will have big dreams, important dreams."

Well, maybe not, I thought. To sleep head to head with a dead lion might empower me—or terrify me. For the time being I decided to simply sit and be quiet in the hut, away from the blazing African sun.

Opening to the Spirits. Each morning before breakfast we shared our dreams with Mandaza. He would then join us in interpreting these dreams. My dreams are usually sparse and unmemorable. My most significant dream came while sleeping in the sacred hut:

My husband and I were invited to a birthday party of an academic acquaintance. First we went to an address we thought was correct, but it wasn't. So we bundled into the car and drove to a hotel reception room filled with many people. We went in and I felt anxious because I didn't recognize anyone and doubted that this was the right party. Eventually we figured it was the wrong party. There was another party in a ballroom upstairs. We trudged up the staircase, more than ready to have something to drink and eat. But this party seemed wrong too. Disappointed we went home.

My friends snickered at my mundane dream report, but Mandaza said: "This is an important dream—it means that you are seeking but not finding. You are having difficulty contacting your ancestors. It is hard for you, but you must be patient and keep working on your dreaming." He was right. I felt humbled by my lack of dreams and this one dream of mine spoke a truth I had not yet accepted. I couldn't seem to break through, and I was feeling frustrated and not able to join in the journey of the mind that I sensed my friends were doing.

(continued)

(continued)

Initiation. One morning additional people had congregated in Mandaza's back yard, mostly patients, some trainees, and one woman who would be initiated as a healer. We were to be initiated as trainees as well. Young men were busy preparing a huge fire outside in the back yard.

Mandaza instructed us initiates to sit in a circle on the lawn. I had not noticed that he had a garden hose in his hand. Suddenly a jet of cold water blasted into my face and I gasped. I felt my heart rate accelerate. While I certainly didn't enjoy the shock, just as certainly, I survived. This was the first of three purifications of the initiation ceremony. Before being able to access the powers of our ancestral spirits we first needed cleansing in body, mind and spirit.

Mandaza fetched a tin can filled with a frothing substance. He stirred this herbal mixture vigorously with a two-pronged stick to create the head of foam. Using a shell for a spoon, we each fed on the bubbles. I had images of vomiting and stomach cramps for all four of us. I was reassured by the knowledge that healers regularly use this brew to promote dreams to enable access their ancestral spirits through their dreams. I hoped it would work for me. Fortunately, the herbal brew had no ill effects. I felt remarkably alert and excited. The day seemed bright—the grass so green and the sky such a deep blue, and everyone was smiling and happy.

Our third purification involved the sweat lodge. After ingesting the foaming bubbles we crawled into the small space, a metal frame covered with tarpaulins. In the middle of the space was a fire made from the red-hot coals and stones from the big outside fire. It was dark and steaming hot inside. Mandaza invited each person to pray to his or her ancestors, thanking them for having brought us from so far to this place in Zimbabwe, and this first level of initiation as healers.

Singing started with several beautiful strong voices leading us in the Shona language. When water was poured onto the hot rocks steam filled the sweat lodge. The heat was almost unbearable and I started to panic. My heart was beating like a train, my face flushed, and my eyes felt so dry I had to shut them. I wanted to escape, but how? There were too many people crammed into this small sweat lodge. I would have to physically crawl over too many hot sweating bodies to find the exit flap.

The singing reached a crescendo of intensity. It filled my head and exalted me. Something about the singing made me feel I was a soaring eagle, flying over the purple mountains and the dry-brown lands of African savanna with the blue winter sky above me. As the heat diminished and the singing grew softer, I felt normal again. I was so relieved I had not made a fool of myself by trying to get out and disturbing this special event for everyone else. I had managed to keep calm in this sauna, and I hate saunas. I was relaxed yet focused, and ready for whatever Mandaza had in mind for us.

(continued)

(continued)

Atonement. The next day Mandaza announced that the strongest and greatest spirits are to be found at the Victoria Falls, alias *Tokaleya Tonga—The Smoke that Thunders.* We drove through grassland with indigenous trees, the *mopanes,* flat-topped thorn trees, and occasional *baobabs*—sentries watching over the lovely land.

After a day of driving we went straight to the Zambezi River, the upper part of the Victoria Falls. A wide swathe of shallow water was gathering momentum just before tumbling 350 ft over the falls to the river below. Our ritual of contrition started.

"First you must make your offerings to the River People. You can throw some coins into the water and I shall watch to see if the River People accept these gifts. Then you must put your head under the water, and keep it under for as long as possible. You must remember the things in your life that you regret, and offer them to the spirits." He joined us in this task, staying under the water for so long that I started worrying about a possible unwelcome encounter with a crocodile.

Everything we did was tinged with potential danger. We were in Africa without touristic safeguards against wild animals, the fast-moving Zambezi River preceding the waterfalls, and the potentially infected river water. While we dried off in the hot sun a young elephant walked by us intent on breaking branches off bushes and eating them. As we were getting back in the car, a herd of buffalo came close, inquisitively watching us. The driver yelled at me to get in the car because these animals can be very dangerous when provoked. But they showed no aggression. And there were no crocodiles either.

"These are good omens for us," declared Mandaza. "The spirits are happy. Now we must travel to Great Zimbabwe, the other place for strong spirits."

Learning and Transforming. Once again we crammed into the car and drove east to Masvinga. Great Zimbabwe is the name of the ruins of an ancient eleventh century African civilization. It consisted of hand built stone walls, about 30–40 ft high. There are passages, towers, and enclosures, but no roof.

While meditating at the eastern Hill Complex section of the ruins, I wondered why I was feeling so blocked. I felt frustrated, even stupid. Then I had an extra-ordinary visualization of a very small crack in a rock that I wanted to go through, but could not. As I contemplated this dilemma a team of grey field mice painlessly dismembered my body, leaving my parts strewn on the rocks around me. This allowed my spiritual self to pass through a small crack in the rocks to a place with a view over the valley. After passing through the crack the field mice reassembled my parts, making a new me. Perhaps some of my western thought commitments were left behind, allowing me to understand and engage in African thinking more easily.

(continued)

(continued)

Accreditation. Later that day our car labored up steep rock slabs with no apparent road. Mandaza had arranged for us to meet with a highly respected, senior healer whose hut perched in an inaccessible place high on the crest of a rocky mountain. She was an old, blind African woman—a famous traditional healer. This diminutive healer greeted us. Then she withdrew into her hut and met with each of us in turn. In the darkness she assessed our respective capacities to be healers. I could only guess that she was conversing with her spirits, and with ours. To my relief, she pronounced me fit to be a healer, without any negative influences hampering me. Perhaps the dismemberment experience rid me of those. This was our final test in the initiation process. I felt elated and properly pleased.

In a follow up ceremony back in Bulawayo, Mandaza and his wife (also a traditional healer) invited me to try out my healing skills on Mandaza himself. Afterwards we prayed together, and Mandaza said I should wear blue beads when I do healings.

I am not a healer by training, and I do not purport to heal anyone in terms of western medicine. My own practice is limited to teaching. Through effective teaching and good communication skills, emotional, and even physical healing can take place. After all, the word 'doctor' comes from the Latin verb *docēre*—to teach.

Fig. 11.1 Spiral aloe (*Aloe polyphylla*) endemic to Lesotho, believed to have healing properties (Photographer: Peter Hewson)

We Can Teach the Children: Additional material to this book can be downloaded from http:// extras.springer.com

Index

© Springer Science+Business Media Dordrecht 2015
M.G. Hewson, *Embracing Indigenous Knowledge in Science and Medical Teaching*,
Cultural Studies of Science Education 10, DOI 10.1007/978-94-017-9300-1

Printed in the United States
By Bookmasters